Introduction to Differentiable Manifolds

LOUIS AUSLANDER
Department of Mathematics
City University of New York

ROBERT E. MACKENZIE
Department of Mathematics
Indiana University

Dover Publications, Inc., New York

To Elinor and Mildred

Published in Canada by General Publishing Company, Ltd., 30 Lesmill Road, Don Mills, Toronto, Ontario.
Published in the United Kingdom by Constable and Company, Ltd., 10 Orange Street, London WC2H 7EG.

This Dover edition, first published in 1977, is an unabridged and corrected republication of the work first published by the McGraw-Hill Book Company, Inc., in 1963.

International Standard Book Number: 0-486-63455-8
Library of Congress Catalog Card Number: 76-55953

Manufactured in the United States of America
Dover Publications, Inc.
180 Varick Street
New York, N.Y. 10014

Preface

This book has as its purpose the study of a mathematical object called a differentiable manifold. Historically, differentiable manifolds arose in many different mathematical disciplines. Such subjects as local differential geometry, projective geometry, algebraic geometry, Riemann surface theory, and continuous or Lie groups have all contributed examples upon which was based the definition of a differentiable manifold. Once the present formulation of the concept of a differentiable manifold was achieved, it was successfully applied to yield extensive generalizations and reformulations of results in the classical disciplines from which it sprang. For instance, the modern treatment of Lie groups could not even be formulated without the language of manifold theory. In addition, differentiable manifolds themselves have served as a source in the development of the mathematical concept of a fiber bundle. Certain studies of manifolds have, in recent years, led to the wholly new subject of differential topology.

The book begins with a leisurely introduction to the general concept of a differentiable manifold. The path that we have chosen to this goal leads through a careful reexamination of the differentiable structure of euclidean space. Once the reader has understood euclidean space as a differentiable manifold, we define the general object and study its elementary properties. We have accompanied our abstract discussion of differentiable manifolds with a liberal dose of some of those historical examples which originally motivated their definition. Once these foundations have been laid, we have selected topics (see the Table of Contents for exact details) which illustrate further the historical background for differentiable manifolds and the directions in which the concepts have been applied. In particular, we have tried to sample the various techniques that have been found useful in handling differentiable manifolds. In a few places, where the technicalities become oppressively intricate or repetitious, the arguments have been left incomplete, and the reader is urged to consult the references at the end of the book for the missing parts. Above all, we hope that the reader will be encouraged by this introduction to pursue the references well beyond the material which is presented here.

This book is not self-contained. The prospective reader will need a solid understanding of the basic theory of finite-dimensional vector spaces and their linear transformations, point-set topology, and that part of analysis which is generally called advanced calculus.

We are, of course, deeply indebted to the previous workers in this field, including the authors listed among the references, whose material we have freely borrowed. And we are especially grateful to Prof. Saunders MacLane for his detailed criticism of former versions of the early parts of this book.

Louis Auslander
Robert E. MacKenzie

Contents

CHAPTER 1

Euclidean, Affine, and Differentiable Structure on R^n

The study of differentiable manifolds begins with the set R^n of all ordered n-tuples $x = (x_1, \ldots, x_n)$, where $x_1, \ldots, x_n$ are real numbers. The set R^n may be endowed with various mathematical structures which the reader has already encountered in geometry, algebra, and analysis. In differential geometry all these structures are called into play simultaneously, and so we shall take the time to distinguish one from another and to study some of the identifications which are customarily made.

1-1 EUCLIDEAN n-SPACE, LINEAR n-SPACE, AND AFFINE n-SPACE

The set R^n may be made into a metric space by defining the distance $d(x, y)$ between the n-tuples $x = (x_1, \ldots, x_n)$ and $y = (y_1, \ldots, y_n)$ by

$$d(x, y) = \sqrt{(x_1 - y_1)^2 + \cdots + (x_n - y_n)^2}$$

This metric space will be called *euclidean n-space* and will be denoted by E^n. A homeomorphism ρ of E^n onto itself which preserves distances, that is, for which $d(\rho(x), \rho(y)) = d(x, y)$, will be called a *rigid motion*. The set of all rigid motions of E^n forms a group which we denote by $R(n)$.

On the other hand, we may introduce into the set R^n the vector space operations

$$c(x_1, \ldots, x_n) = (cx_1, \ldots, cx_n)$$
$$(x_1, \ldots, x_n) + (y_1, \ldots, y_n) = (x_1 + y_1, \ldots, x_n + y_n)$$

This vector space may be given the topology of the n-fold topological product of the real line with itself. With respect to this topology, the set R^n becomes a topological vector space; that is, the mapping of $R^1 \times R^n$ into R^n given by $(c, x) \to cx$ and the mapping of $R^n \times R^n$ into

R^n given by $(x, y) \to x + y$ are continuous. This topological vector space will be called *linear n-space* and will be denoted by V^n. A homeomorphism of V^n onto itself which preserves the operations is, of course, the same thing as a nonsingular linear transformation. The set of all nonsingular linear transformations of V^n forms a group called the general linear group and denoted by $GL(n)$.

The reader should note that we have not assumed the existence of a metric on V^n, although the topology that we have put on V^n is equivalent to the topology on E^n. Indeed, our object is to investigate the relationship between the metric on E^n and the algebraic operations in V^n.

If in E^n we retain the topology, the notion of straight line, and the notion of parallelism among straight lines but discard the metric structure, we obtain *affine n-space*, which we shall denote by A^n. In A^n we may speak of continuous mappings or of parallel lines, but we cannot speak of the distance between points. A homeomorphism of A^n that maps lines onto lines is called an *affinity*. The set of all affinities of A^n forms a group which we denote by $A(n)$.

Let $x \in E^n$. A homeomorphism τ_x: $V^n \to E^n$ of V^n onto E^n for which $\tau_x(0) = x$ will be said to *attach* V^n to the point x. Such a homeomorphism enables the additive group of V^n to act on E^n as a group of homeomorphisms. That is to say, if $u \in V^n$, then ρ_u: $E^n \to E^n$ is a homeomorphism of E^n defined by

$$\rho_u(y) = \tau_x(\tau_x^{-1}(y) + u)$$

and the homeomorphisms ρ_u satisfy the rule

$$\rho_u \circ \rho_v = \rho_{u+v}$$

where $\rho_u \circ \rho_v$ is the composition of the mappings ρ_u and ρ_v. In a similar fashion, if $x \in A^n$, a homeomorphism τ_x: $V^n \to A^n$ attaches V^n to the point x and enables the additive group of V^n to act on A^n.

Lemma 1-1. *Let V^n be attached to E^n by means of the homeomorphism τ_x: $V^n \to E^n$. Then the elements of V^n act on E^n as rigid motions if and only if they act on E^n so as to move each line onto a parallel line.*

Proof. 1. Assume that the elements of V^n act on E^n as rigid motions. Let $y \in E^n$ and $u \in V^n$. Let $v = \tau_x^{-1}(y)$. Then

$$\rho_u(x) = \tau_x(u)$$

and so $\rho_v(\rho_u(x)) = \tau_x(u + v)$ On the other hand, $\rho_u(y) = \tau_x(u + v)$, and so $\rho_v(\rho_u(x)) = \rho_u(y)$. Since $\rho_v(x) = y$, we find that ρ_v moves the points x and $\rho_u(x)$ onto the points y and $\rho_u(y)$, respectively. Since ρ_v

is a rigid motion, we conclude that $d(y, \rho_u(y)) = d(x, \rho_u(x))$ for all $y \in E^n$; that is, ρ_u moves all points of E^n the same distance.

The line through the points y^1 and y^2 in E^n may be characterized as follows by means of the metric: z lies on the line through y^1 and y^2 if and only if one of the following relations holds:

$$\begin{aligned} d(z, y^1) + d(y^1, y^2) &= d(z, y^2) \\ d(y^1, z) + d(z, y^2) &= d(y^1, y^2) \\ d(y^1, y^2) + d(y^2, z) &= d(y^1, z) \end{aligned}$$

Since ρ_u is a rigid motion, these relations persist when y^1, y^2, z are replaced by $\rho_u(y^1)$, $\rho_u(y^2)$, $\rho_u(z)$, respectively. Consequently ρ_u carries lines of E^n onto lines.

Moreover, two lines in E^n are parallel if and only if the distances from one of the lines to the points of the other are bounded. It follows that ρ_u carries each line in E^n onto another which is parallel to it.

2. Conversely, assume that the elements of V^n act on E^n so as to move each line onto a parallel line. As in the proof of part 1, let $y \in E^n$, $u \in V^n$, and $v = \tau_x^{-1}(y)$. Then ρ_v moves the points x and $\rho_u(x)$ onto the points y and $\rho_u(y)$, respectively. Hence the line through y and $\rho_u(y)$ is parallel to the line through x and $\rho_u(x)$. But also the line through $\rho_u(x)$ and $\rho_u(y)$ is parallel to the line through x and y. Hence x, y, $\rho_u(y)$, and $\rho_u(x)$ are the vertices of a parallelogram, and it follows that $d(\rho_u(x), \rho_u(y)) = d(x, y)$.

If y, z are two points in E^n, the triangle with vertices $\rho_u(x)$, $\rho_u(y)$, and $\rho_u(z)$ is similar to the triangle with vertices x, y, and z. Since the length of one side of the first triangle is equal to the length of the corresponding side of the other, they are actually congruent. Hence $d(\rho_u(y), \rho_u(z)) = d(y, z)$, which says that ρ_u is a rigid motion.

If σ_x: $V^n \to E^n$ is a homeomorphism which attaches V^n to E^n in such a way that the elements of V^n act on E^n as rigid motions, these rigid motions are called *translations* of E^n. If σ_x: $V^n \to A^n$ attaches V^n to A^n in such a way that the elements of V^n act on A^n so as to carry each line onto a parallel line, these affinities are called *translations* of A^n. Lemma 1-1 expresses the fact that translations in E^n are equivalent to translations in A^n.

We shall now see that the translations of A^n (or of E^n) are defined independently of the way in which V^n is attached to a point of A^n (or of E^n).

Theorem 1-1. *If $x \in A^n$, then it is possible to attach V^n to A^n at x by a homeomorphism σ_x: $V^n \to A^n$ in such a way that the elements of V^n act on A^n as translations. If τ_x: $V^n \to A^n$ is a second homeomorphism* *which has this property, then* $\sigma_x^{-1} \circ \tau_x \colon V^n \to V^n$ *is a nonsingular linear transformation.*

Proof. Let $\{v^1, \ldots, v^n\}$ be a basis of V^n. Define the mapping $\sigma_x \colon V^n \to A^n$ as follows: If $u = c_1v^1 + \cdots + c_nv^n$, then

$$\sigma_x(u) = (c_1 + x_1, \ldots, c_n + x_n)$$

Then

$$\sigma_x(\sigma_x^{-1}(y) + u) = (c_1 + y_1, \ldots, c_n + y_n)$$

so that the elements of V^n act on A^n as translations.

Let $\tau_x \colon V^n \to A^n$ be a homeomorphism which attaches V^n to the point x of A^n in such a way that for each $u \in V^n$ the mapping $\rho_u \colon A^n \to A^n$ given by

$$\rho_u(y) = \tau_x(\tau_x^{-1}(y) + u)$$

is a translation. Let $\lambda = \sigma_x^{-1} \circ \tau_x$. Let $u, v \in V^n$, and let $y = \tau_x(u)$. Then $\rho_v(x) = \tau_x(v)$ and $\rho_v(y) = \tau_x(u + v)$. It follows that the line through $\tau_x(0)$ and $\tau_x(u)$ is parallel to the line through $\tau_x(v)$ and $\tau_x(u + v)$. Similarly, the line through $\tau_x(0)$ and $\tau_x(v)$ is parallel to the line through $\tau_x(u)$ and $\tau_x(u + v)$. Consequently, $\tau_x(0)$, $\tau_x(u)$, $\tau_x(u + v)$, and $\tau_x(v)$ are the vertices of a parallelogram in A^n. If we apply σ_x^{-1}, we obtain

$$\lambda(u + v) = \lambda(u) + \lambda(v)$$

In particular, we may conclude that $\lambda(mu) = m\lambda(u)$ for any positive integer m. If we replace mu by u, we also obtain

$$\lambda\left(\left(\frac{1}{m}\right)u\right) = \left(\frac{1}{m}\right)\lambda(u)$$

Thus $\lambda(ru) = r\lambda(u)$ for every positive rational number r. From $\lambda(0) = 2\lambda(0)$ we find that $\lambda(0) = 0$ and hence that $\lambda(-u) = -\lambda(u)$. Combining these results, we obtain $\lambda(ru) = r\lambda(u)$ for every rational number r. Since λ is continuous, $\lambda(cu) = c\lambda(u)$ for every real number c. Thus λ is a linear transformation. Since λ is one-to-one, it must be nonsingular.

Actually we have proved a bit more than is indicated in the statement of Theorem 1-1. For we have given a formula for the homeomorphism σ_x.

By virtue of Lemma 1-1, we may replace A^n by E^n in the statement of Theorem 1-1.

Theorem 1-2. *If* $x \in E^n$, *then it is possible to attach* V^n *to* E^n *at* x *by a homeomorphism* $\sigma_x \colon V^n \to E^n$ *in such a way that the elements of* V^n *act on* E^n *as rigid motions. If* $\tau_x \colon V^n \to E^n$ *is a second homeomorphism* *which has this property, then* $\sigma_x^{-1}\circ\tau_x$: $V^n \to V^n$ *is a nonsingular linear transformation.*

Let σ_x: $V^n \to E^n$ be a homeomorphism which attaches V^n to the point x in E^n. If σ_x is such that the elements of V^n act on E^n as translations, we shall say that σ_x attaches V^n *affinely* to E^n. Let λ: $V^n \to V^n$ be a nonsingular linear transformation, and let $\tau_x = \sigma_x\circ\lambda$. Then

$$\tau_x(\tau_x^{-1}(y) + u) = \sigma_x(\sigma_x^{-1}(y) + \lambda(u))$$

We see from this that τ_x attaches V^n to E^n affinely if and only if σ_x does, and when this is true, both determine the same family of translations of E^n. When $\tau_x = \sigma_x\circ\lambda$, we shall say that τ_x and σ_x are *equivalent.* Theorem 1-2 tells us that there is only one equivalence class of homeomorphisms which attach V^n affinely to the point x in E^n. Consequently, the family of translations of E^n is unique. The explicit formula for $\sigma_x(\sigma_x^{-1}(y) + u)$ given in the proof of Theorem 1-1 shows that this family is independent of the point x.

Theorem 1-3. *Let* V^n *be attached affinely to each point of* A^n *by means of the family* $\{\sigma_x\}$ *of homeomorphisms. If* ρ: $A^n \to A^n$ *is an affinity, then* $\sigma_{\rho(x)}^{-1}\circ\rho\circ\sigma_x$: $V^n \to V^n$ *is a nonsingular linear transformation for each* $x \in A^n$.

Proof. Let $\lambda = \sigma_{\rho(x)}^{-1}\circ\rho\circ\sigma_x$. Let l_1 and l_2 be two distinct lines through a point y. Then $\rho(l_1)$ and $\rho(l_2)$ must be two distinct lines through the point $\rho(y)$. We may describe the plane P determined by a pair of lines l_1 and l_2 as the set of points which lie on those lines that intersect l_1 and l_2 in distinct points, together with the point y. Thus $\rho(P)$ is the plane determined by $\rho(l_1)$ and $\rho(l_2)$. We have therefore proved that ρ maps planes onto planes. It follows that ρ maps parallel lines onto parallel lines.

Let u, $v \in V^n$. According to Theorem 1-1, which gives us a formula for computing σ_x, the line l_1 through $\sigma_x(0)$ and $\sigma_x(u)$ must be parallel to the line l_1' through $\sigma_x(v)$ and $\sigma_x(u + v)$. Similarly, the line l_2 through $\sigma_x(0)$ and $\sigma_x(v)$ must be parallel to the line l_2' through $\sigma_x(u)$ and $\sigma_x(u + v)$. Since the pairs of lines $\rho(l_1)$, $\rho(l_1')$ and $\rho(l_2)$, $\rho(l_2')$ are parallel, it follows that $\rho\circ\sigma_x(0)$, $\rho\circ\sigma_x(u)$, $\rho\circ\sigma_x(u + v)$, and $\rho\circ\sigma_x(v)$ are the vertices of a parallelogram, provided that these four points are not collinear. Applying $\sigma_{\rho(x)}^{-1}$, we obtain $\lambda(u + v) = \lambda(u) + \lambda(v)$. The provision that the four points are not collinear is equivalent to the condition that u and v are linearly independent. However, λ is a continuous mapping so that the relation

$$\lambda(u + v) = \lambda(u) + \lambda(v)$$

holds generally. It now follows, as in the proof of Theorem 1-1, that λ is a nonsingular linear transformation.

Since a rigid motion of E^n carries straight lines into straight lines, we may replace A^n by E^n in Theorem 1-3.

Theorem 1-4. *Let V^n be attached affinely to each point of E^n by means of the family $\{\sigma_x\}$ of homeomorphisms. If $\rho: E^n \to E^n$ is a rigid motion, then $\sigma_{\rho(x)}^{-1} \circ \rho \circ \sigma_x: V^n \to V^n$ is a nonsingular linear transformation for each $x \in E^n$.*

To summarize briefly, Theorem 1-4 shows that it is possible to attach V^n to each point of E^n in such a way that a rigid motion of E^n induces nonsingular linear transformations of V^n at each point. If we insist that V^n be attached affinely at each point, Theorem 1-2 shows that this can be done in essentially only one way, if we identify equivalent homeomorphisms.

However, we are chiefly concerned with the situation where the set R^n is provided with a differentiable structure. Briefly, this means that R^n is not only a topological space but that among the continuous real-valued functions on R^n we distinguish certain ones called the differentiable functions. A homeomorphism of R^n that preserves the class of differentiable functions will be called a diffeomorphism. In this situation we shall consider the problem of how to "attach" V^n to each point of R^n in such a way that the diffeomorphisms of R^n will induce nonsingular linear transformations of V^n at each point. Although this cannot be done, as it was in this section, by means of homeomorphisms $\sigma_x: V^n \to R^n$, we shall see that the problem has a satisfactory solution.

Our interest in "attaching" V^n to R^n lies in the expectation that significant properties of mappings of R^n will be reflected in the mappings which are induced on V^n. For rigid motions in E^n or affinities in A^n the induced mappings reflect all the properties. However, for differentiable mappings, only the behavior in the neighborhood of a point is reflected, as we shall see.

1-2 DIFFERENTIABLE FUNCTIONS AND MAPPINGS ON LINEAR n-SPACE

If $\sigma_x: V^n \to A^n$ is a homeomorphism that attaches V^n to the point x, we may use σ_x to transfer the affine structure of A^n to V^n or to transfer the algebraic structure of V^n to A^n. We shall simply say that we have used $\sigma_x: V^n \to A^n$ to *identify* A^n with V^n. In a similar fashion, we may identify E^n with V^n so as to put a metric on V^n.

Let E^n be identified with V^n by means of a homeomorphism $\sigma_x: V^n \to E^n$ which attaches V^n affinely to the point x. If $v \in V^n$, the numbers $y_1, \ldots, y_n$ determined by the relation $\sigma_x(v) = (y_1, \ldots, y_n)$ will be called the *coordinates* of v. If f is a real-valued function defined on an open set U in V^n, the composite mapping $f \circ \sigma_x^{-1}$ is a real-

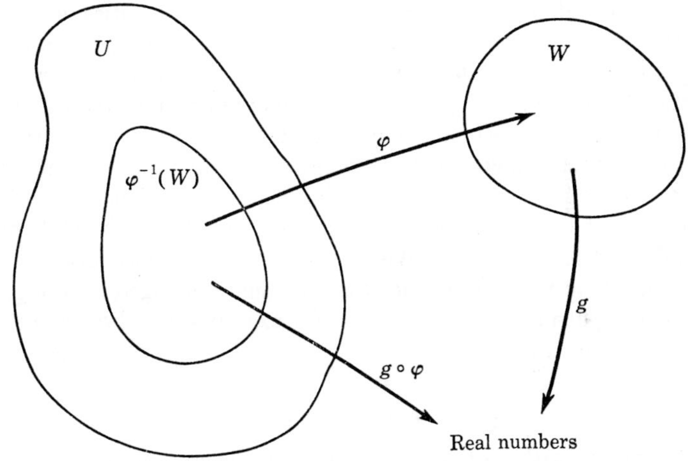

FIG. 1-1

valued function of the variables $y_1, \ldots, y_n$ on an open set in E^n. We shall say that f is *differentiable* on U if $f \circ \sigma_x^{-1}$ has partial derivatives of all orders with respect to the coordinates $y_1, \ldots, y_n$ at each point of $\sigma_x(U)$. It is easy to see that this notion of differentiability does not depend on the point x in E^n or the choice of the homeomorphism $\sigma_x: V^n \to E^n$ which attaches V^n to the point x.

Let U be an open subset of V^n. Consider a mapping $\varphi: U \to V^m$. The mapping φ is said to be *differentiable* on U if for every function g which is differentiable on an open subset W of V^m the composite function $g \circ \varphi$ is differentiable on $\varphi^{-1}(W)$ (see Figure 1-1). In case the set $\varphi^{-1}(W)$ is empty, this condition is vacuously satisfied. Let $(z_1, \ldots, z_m)$ be coordinates for the points of V^m. In particular, we may write $(z_1(v), \ldots, z_m(v))$ for the coordinates of the point $\varphi(v)$ if $v \in U$. Then $z_1(v), \ldots, z_m(v)$ are real-valued functions on U, and the differentiability of the mapping φ on U is equivalent to the simultaneous differentiability of the functions $z_1(v), \ldots, z_m(v)$ on U.

Let $\varphi: V^n \to V^n$ be a one-to-one differentiable mapping onto V^n. If the inverse mapping $\varphi^{-1}: V^n \to V^n$ is also differentiable, φ is called a *diffeomorphism* of V^n. That the inverse of a one-to-one differentiable mapping need not be differentiable can be seen in the case $n = 1$. Consider $y = x^3$, whose inverse $x = y^{1/3}$ is not differentiable on any neighborhood of 0. If φ_1 and φ_2 are diffeomorphisms of V^n, then so is the composite mapping $\varphi_1 \circ \varphi_2$. Consequently, the set of all diffeomorphisms of V^n forms a group.

If we identify A^n with V^n and also E^n with V^n by attaching V^n to these spaces affinely, we are able to consider affinities and rigid motions of V^n. These are easily seen to be diffeomorphisms of V^n. Thus the group of diffeomorphisms of V^n contains the group of affinities, and this group, in turn, contains the group of rigid motions. Properties, such as the distance between points and the angles between lines, that remain unchanged by the group of rigid motions are called *metric invariants.* Those, such as the collinearity of points and the parallelism of lines, that remain unchanged by the group of affinities are called *affine invariants.* We may similarly define the *differentiable invariants* to be those properties that remain unchanged by the group of diffeomorphisms. If we consider, for example, the diffeomorphism $\varphi: V^2 \to V^2$ given by

$$\varphi: \quad \begin{aligned} y_1 &= x_1 \\ y_2 &= x_1^2 + x_2 \end{aligned}$$

we can see that the affine structure of V^n is not preserved by the diffeomorphisms of V^n. It is therefore apparent that the group of diffeomorphisms is larger than the group of affinities, and so affine invariants are not generally differentiable invariants. One says that the differentiable structure of V^n is "weaker" than the affine structure.

EXERCISES

1. Let V^n be attached affinely to x in A^n by means of a homeomorphism $\sigma_x: V^n \to A^n$. Show that the vector space operations which are introduced into E^n by this identification are given by

$$(y_1, \ldots, y_n) + (z_1, \ldots, z_n) = (y_1 + z_1 - x_1, \ldots, y_n + z_n - x_n)$$
$$c(y_1, \ldots, y_n) = (cy_1 + (1 - c)x_1, \ldots, cy_n + (1 - c)x_n)$$

2. Prove that the differentiability of a function on an open subset of V^n does not depend on the choice of the homeomorphism by which we attach V^n affinely to E^n.

3. If $(z_1, \ldots, z_m)$ are coordinates in V^m, if $\varphi: U \to V^m$ is a mapping of the open set U in V^n into V^m, and if $(z_1(v), \ldots, z_m(v))$ are the coordinates of $\varphi(v)$, prove that $z_1(v), \ldots, z_m(v)$ are differentiable functions on U if and only if φ is differentiable on U.

4. Let $\varphi: W \to U$, where W is an open subset of V^n and U is an open subset of V^m, and let $\psi: U \to V^l$. Prove that, if φ and ψ are differentiable, the composite mapping $\psi \circ \varphi: W \to V^l$ is differentiable.

5. Consider the mapping $\varphi: V^2 \to V^2$ given by

$$\varphi: \qquad \begin{aligned} y_1 &= x_1 \cos x_2 \\ y_2 &= x_1 \sin x_2 \end{aligned}$$

Show that φ is one-to-one on a sufficiently small neighborhood of each point (x_1, x_2) of V^2 with $x_1 \neq 0$.

1-3 THE TANGENT SPACE AND COTANGENT SPACE AT A POINT OF V^n

Let J denote the open interval $-1 < t < 1$ and $\bar{J}$ the closed interval $-1 \leqq t \leqq 1$. A continuous mapping $\gamma: \bar{J} \to V^n$ is called a *parametrized curve* in V^n. The image of the point t of $\bar{J}$ will be denoted by $\gamma(t)$. If $v = \gamma(0)$, then γ will be said to be a curve *through* v. If the restriction of γ to the open interval J is a differentiable mapping, γ will be called a *differentiable* parametrized curve.

Let v be a point of V^n. Let $\mathfrak{C}$ be the set of all differentiable parametrized curves through v. Let $\mathfrak{F}$ be the set of all functions that are defined and differentiable near v; that is, if $f \in \mathfrak{F}$, there is a neighborhood U of v on which f is differentiable. If $\gamma \in \mathfrak{C}$ and $f \in \mathfrak{F}$, then $f \circ \gamma$ is a real-valued differentiable function of t, provided that t is sufficiently small. We may therefore define

$$\langle \gamma, f \rangle = \frac{d}{dt} f \circ \gamma \bigg|_{t=0}$$

The "product" $\langle \gamma, f \rangle$ gives rise to an equivalence relation in $\mathfrak{C}$ and also an equivalence relation in $\mathfrak{F}$ as follows: We say that two curves γ_1 and γ_2 from $\mathfrak{C}$ are equivalent when $\langle \gamma_1, f \rangle = \langle \gamma_2, f \rangle$ for all f in $\mathfrak{F}$. Similarly, two functions f_1 and f_2 from $\mathfrak{F}$ are equivalent when $\langle \gamma, f_1 \rangle = \langle \gamma, f_2 \rangle$ for all curves γ in $\mathfrak{C}$.

The equivalence classes of curves of $\mathfrak{C}$ will be called *tangent vectors* to V^n at v. We shall presently show that the set of all tangent vectors to V^n at v may be given the structure of a vector space. This space will be called the *tangent space* to V^n at v and will be denoted by $T(v)$. Geometrically, a tangent vector to V^n at v may be thought of as a family of curves, all of which are tangent to one another at v (see Figure 1-2). If γ is a curve in the equivalence class α, we shall say that α is the tangent vector to γ at v.

Presently we shall see that $T(v)$ is a vector space of dimension n. It is as the tangent space $T(v)$ that we "attach" V^n to the point v.

The reader should recognize that we are engaged in solving the problem described at the end of Section 1-1.

The equivalence classes of functions from the set $\mathfrak{F}$ will be called *differentials* at v. We shall see that the set of all differentials at v may also be given the structure of a vector space. This space will be called the *cotangent space* to V^n at v and will be denoted by $T^*(v)$. If f is a function from $\mathfrak{F}$, the equivalence class to which f belongs will be called the *differential of* f and will be denoted by df.

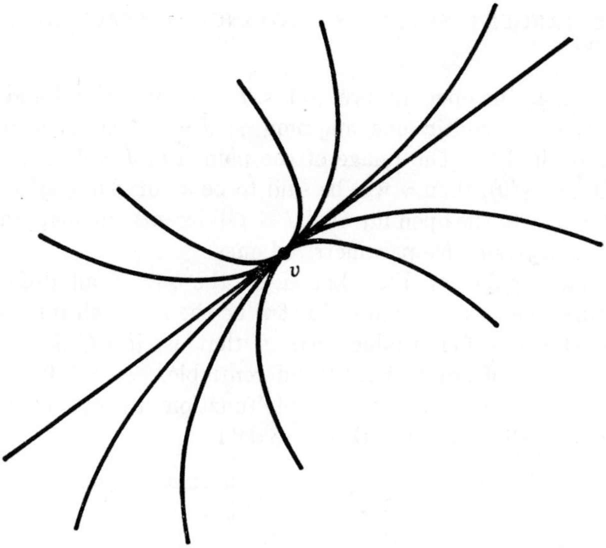

FIG. 1-2

If α is the tangent vector to the curve γ in $\mathfrak{C}$ and f is a function from $\mathfrak{F}$, we shall define the value of df at α, denoted by $\langle \alpha, df \rangle$, by $\langle \alpha, df \rangle = \langle \gamma, f \rangle$. We shall find that $\langle \alpha, df \rangle$ is a product in the algebraic sense; that is, it is linear in each of its two arguments.

In order to facilitate the introduction of vector space operations into $T^*(v)$, we shall employ another equivalence relation among the functions of $\mathfrak{F}$. It is clear that, in order to compute the value of $\langle \gamma, f \rangle$, we need only know the values of f on a neighborhood of v. Consequently, for this purpose we shall cease to distinguish between two functions that have the same values on a neighborhood of v. The equivalence classes in $\mathfrak{F}$ which arise from this equivalence relation will be called the *germs* of the functions of $\mathfrak{F}$. The set of all germs will be denoted by

$G^*(v)$. If f is a function which belongs to the germ κ, we may define $\langle \gamma, \kappa \rangle = \langle \gamma, f \rangle$.

The operations among functions, which are already at hand, may now be used to introduce vector space operations in $G^*(v)$. Let f_1 and f_2 be functions from $\mathfrak{F}$ and let κ_1 and κ_2 be their germs, respectively. Let a_1 and a_2 be two numbers. There is a neighborhood of v on which both f_1 and f_2 are defined and differentiable and hence on which the function $a_1f_1 + a_2f_2$ is defined and differentiable. We shall define $a_1\kappa_1 + a_2\kappa_2$ to be the germ of the function $a_1f_1 + a_2f_2$. Then $G^*(v)$ becomes a vector space.

If f_1 and f_2 are functions with the same germ, then $df_1 = df_2$. Hence we may map $G^*(v)$ into $T^*(v)$ in the following way: Let f be any function in the germ κ and map κ onto df. We may also speak of df as the differential of κ and denote it by $d\kappa$. Let Z^* denote the subspace of $G^*(v)$ consisting of those germs whose differentials are zero. Since $d\kappa_1 = d\kappa_2$ if and only if $\kappa_1 - \kappa_2$ has the differential zero, it follows that the mapping of $G^*(v)/Z^*$ onto $T^*(v)$ which is given by

$$\kappa + Z^* \to d\kappa$$

is well-defined and one-to-one. Consequently we may give to $T^*(v)$ the vector space structure of $G^*(v)/Z^*$ by defining

$$a_1\, d\kappa_1 + a_2\, d\kappa_2 = d(a_1\kappa_1 + a_2\kappa_2)$$

Since differentiation is a linear operator, we also obtain the rule

$$\langle \alpha, a_1\kappa_1 + a_2\kappa_2 \rangle = a_1\langle \alpha, \kappa_1 \rangle + a_2\langle \alpha, \kappa_2 \rangle$$

Hence if λ_1 and λ_2 are elements of $T^*(v)$, then

$$\langle \alpha, a_1\lambda_1 + a_2\lambda_2 \rangle = a_1\langle \alpha, \lambda_1 \rangle + a_2\langle \alpha, \lambda_2 \rangle$$

The mapping $\lambda \to \langle \alpha, \lambda \rangle$ of $T^*(v)$ into the real numbers, which we may associate with the tangent vector α, is a linear functional on $T^*(v)$. Each tangent vector α gives rise to a distinct linear functional, and so we may regard $T(v)$ as a subset of the dual space $T^*(v)^*$ to $T^*(v)$.

Lemma 1-2. *$T(v)$ is a subspace of $T^*(v)^*$.*

Proof. Let α_1 and α_2 be two tangent vectors to V^n at v. Let $\varphi_1(\lambda) = \langle \alpha_1, \lambda \rangle$ and $\varphi_2(\lambda) = \langle \alpha_2, \lambda \rangle$ for all λ in $T^*(v)$. Let $\varphi = a_1\varphi_1 + a_2\varphi_2$ in $T^*(v)^*$, where a_1 and a_2 are numbers. We must show that there is a tangent vector α such that $\varphi(\lambda) = \langle \alpha, \lambda \rangle$ for all λ in $T^*(v)$.

Let γ_1 and γ_2 be differentiable parametrized curves through v with the tangent vectors α_1 and α_2, respectively. Define the curve γ by

$$\gamma(t) = a_1\gamma_1(t) + a_2\gamma_2(t) + (1 - a_1 - a_2)v$$

for $-1 \leqq t \leqq 1$. Then γ is in $\mathfrak{C}$. If $f \in \mathfrak{F}$, then

$$\frac{d}{dt}f[a_1\gamma_1(t) + a_2\gamma_2(t) + (1 - a_1 - a_2)v] = a_1\frac{d}{dt}f[\gamma_1(t)] + a_2\frac{d}{dt}f[\gamma_2(t)]$$

and so

$$\langle\gamma, f\rangle = a_1\langle\gamma_1, f\rangle + a_2\langle\gamma_2, f\rangle$$

It follows that, if α is the tangent vector to γ at v, then

$$\langle\alpha, \lambda\rangle = a_1\langle\alpha_1, \lambda\rangle + a_2\langle\alpha_2, \lambda\rangle$$

for all λ in $T^*(v)$, which was to be proved.

We see from the preceding lemma that we may give to $T(v)$ the vector space structure that it acquires as a subspace of $T^*(v)^*$. We then have the rule

$$\langle a_1\alpha_1 + a_2\alpha_2, \lambda\rangle = a_1\langle\alpha_1, \lambda\rangle + a_2\langle\alpha_2, \lambda\rangle$$

Let $(x_1, \ldots, x_n)$ be coordinates in V^n. Let w be the vector with the coordinates $(0, \ldots, 0)$, and let $u^j(t)$ be the vector with the coordinates $(0, \ldots, t, \ldots, 0)$, where the t occurs in the jth place. Then define the curve γ_j by

$$\gamma_j(t) = v + u^j(t) - w$$

for $-1 \leqq t \leqq 1$. We shall denote the tangent vector to γ_j at v by $(\partial/\partial x_j)_v$ or by $\partial/\partial x_j$ if it is clear at which point this tangent vector is being considered. Define the function f_j by $f_j(x_1, \ldots, x_n) = x_j$. The differential of f_j at v will be denoted by $(dx_j)_v$ or by dx_j when we wish to suppress the point v. If v has the coordinates $(a_1, \ldots, a_n)$, then

$$f_j(\gamma_i(t)) = \begin{cases} a_j & \text{if } i \neq j \\ a_j + t & \text{if } i = j \end{cases}$$

Hence

$$\left\langle\frac{\partial}{\partial x_i}, dx_j\right\rangle = \begin{cases} 0 & \text{if } i \neq j \\ 1 & \text{if } i = j \end{cases} \tag{1-3-1}$$

Theorem 1-5. *Let $(x_1, \ldots, x_n)$ be coordinates in V^n. Then the vector spaces $T(v)$ and $T^*(v)$ are dual to one another and of dimension n. Furthermore, $\{\partial/\partial x_1, \ldots, \partial/\partial x_n\}$ is a basis of $T(v)$ and $\{dx_1, \ldots, dx_n\}$ is its dual basis.*

Proof. From the relations (1-3-1) it is clear that the differentials $dx_1, \ldots, dx_n$ are linearly independent. We shall prove that they span $T^*(v)$. If γ is any curve in $\mathfrak{C}$, there are functions $g_1, \ldots, g_n$ of t such that $\gamma(t)$ is the point whose coordinates are $(g_1(t), \ldots, g_n(t))$. If f is a function from $\mathfrak{F}$, then

$$\langle \gamma, df \rangle = \sum_{j=1}^{n} \left.\frac{\partial f}{\partial x_j}\right|_v g_j'(0)$$

On the other hand,

$$\langle \gamma, dx_j \rangle = g_j'(0)$$

and so

$$\langle \gamma, df \rangle = \left\langle \gamma, \sum_{j=1}^{n} \left.\frac{\partial f}{\partial x_j}\right|_v dx_j \right\rangle$$

It follows that

$$df = \sum_{j=1}^{n} \left.\frac{\partial f}{\partial x_j}\right|_v dx_j$$

This proves that $T^*(v)$ has dimension n.

The dual space $T^*(v)^*$ to $T^*(v)$ therefore has dimension n. Since $T(v)$ is a subspace of $T^*(v)^*$, its dimension cannot exceed n. On the other hand, from the relations (1-3-1) we may conclude that the tangent vectors $\partial/\partial x_1, \ldots, \partial/\partial x_n$ are linearly independent. Consequently $\dim T(v) = n$ and hence $T(v) = T^*(v)^*$. The duality of the two bases is expressed by the relations (1-3-1).

Let $\varphi\colon U \to V^m$ be a differentiable mapping of an open subset U of V^n which contains the point v. If f is a function which is differentiable on a neighborhood of the point $\varphi(v)$, then $f\circ\varphi$ is in $\mathfrak{F}$ (see Figure 1-3). Furthermore, if f_1 and f_2 are two such functions which belong to the same germ at $\varphi(v)$, then $f_1\circ\varphi$ and $f_2\circ\varphi$ belong to the same germ at v. Consequently we may define

$$\varphi^{\#}\,(\text{germ of } f \text{ at } \varphi(v)) = \text{germ of } f\circ\varphi \text{ at } v$$

and obtain a linear transformation $\varphi^{\#}: G^*(\varphi(v)) \to G^*(v)$.

On the other hand, if γ is a curve in $\mathfrak{C}$, then $\varphi\circ\gamma$ is a curve in V^m through $\varphi(v)$ (see Figure 1-4). Since $f\circ(\varphi\circ\gamma) = (f\circ\varphi)\circ\gamma$ for functions f defined near $\varphi(v)$ and curves γ in $\mathfrak{C}$, we see that

$$\langle \varphi\circ\gamma, d\kappa \rangle = \langle \gamma, d\varphi^{\#}(\kappa) \rangle \tag{1-3-2}$$

where κ is the germ of f.

If κ_1 and κ_2 are two germs at $\varphi(v)$ with $d\kappa_1 = d\kappa_2$, then, from (1-3-2),

$$\langle \gamma, d\varphi^{\#}(\kappa_1) \rangle = \langle \gamma, d\varphi^{\#}(\kappa_2) \rangle$$

for all curves γ in $\mathfrak{C}$. Hence $d\varphi^{\#}(\kappa_1) = d\varphi^{\#}(\kappa_2)$. We may therefore define $\varphi^*(d\kappa) = d\varphi^{\#}(\kappa)$ and obtain a linear transformation

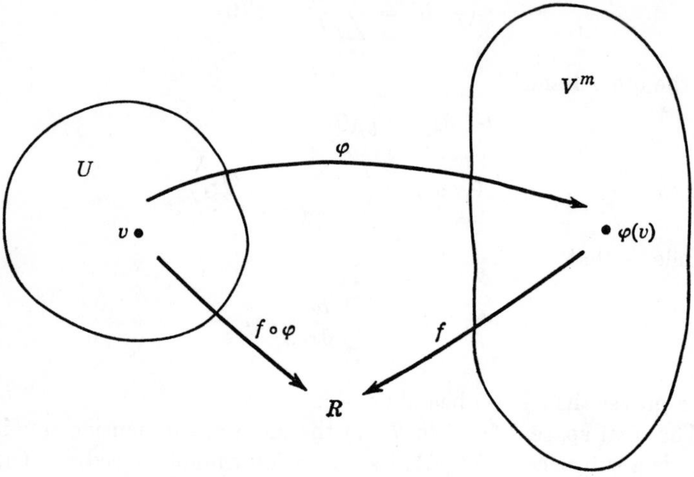

FIG. 1-3

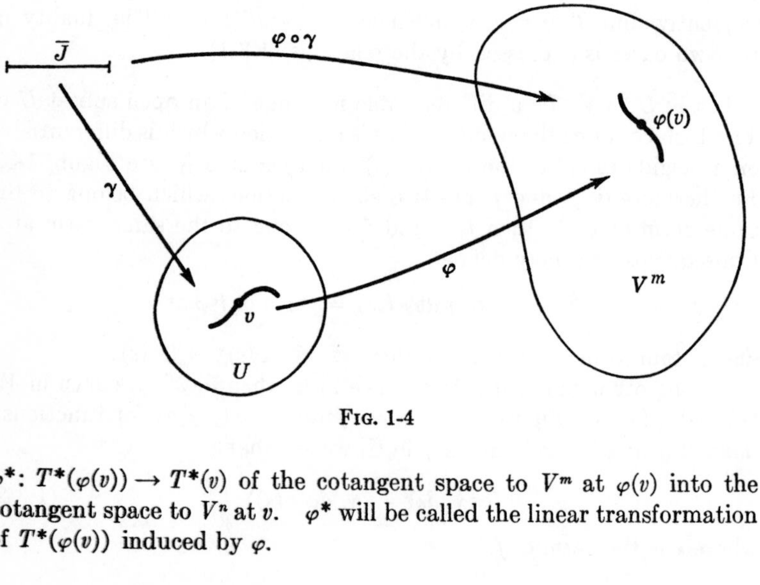

FIG. 1-4

$\varphi^*\colon T^*(\varphi(v)) \to T^*(v)$ of the cotangent space to V^m at $\varphi(v)$ into the cotangent space to V^n at v. φ^* will be called the linear transformation of $T^*(\varphi(v))$ induced by φ.

Similarly, if γ_1 and γ_2 are two curves in $\mathfrak{C}$ with the same tangent vector, then

$$\langle \varphi \circ \gamma_1, d\kappa \rangle = \langle \varphi \circ \gamma_2, d\kappa \rangle$$

for all germs κ at $\varphi(v)$. Hence $\varphi \circ \gamma_1$ and $\varphi \circ \gamma_2$ have the same tangent vector at $\varphi(v)$. We may define

$$\varphi_*(\text{tangent vector to } \gamma) = \text{tangent vector to } \varphi \circ \gamma$$

and obtain a mapping $\varphi_*: T(v) \to T(\varphi(v))$ of the tangent space to V^n at v into the tangent space to V^m at $\varphi(v)$. The mappings φ^* and φ_* satisfy the relation

$$\langle \varphi_*(\alpha), \lambda \rangle = \langle \alpha, \varphi^*(\lambda) \rangle \tag{1-3-3}$$

for $\alpha \in T(v)$ and $\lambda \in T^*(\varphi(v))$. Since φ^* is a linear transformation, it follows from the relation (1-3-3) that φ_* is a linear transformation. φ_* will be called the linear transformation of $T(v)$ induced by φ.

EXERCISES

1. Let $(x_1, \ldots, x_n)$ be coordinates in V^n. Let γ be a parametrized curve in V^n. Let $(g_1(t), \ldots, g_n(t))$ be the coordinates of the point $\gamma(t)$. Prove that γ is differentiable if and only if $g_1, \ldots, g_n$ are differentiable.

2. Prove that, if γ_1 and γ_2 are two curves in $\mathfrak{C}$ with the same tangent vector at v, then γ_1 and γ_2 are tangent to each other at v.

3. Construct an example of two differentiable curves in V^2 which are tangent to one another at 0 and yet do not have the same tangent vector.

4. Consider the functions

$$f_1(t) = \begin{cases} 0 & \text{for } t \leqq 1 \\ e^{-1/(t-1)^2} & \text{for } t > 1 \end{cases}$$
$$f_2(t) = 0 \qquad \text{for } -\infty < t < \infty$$

Prove that f_1 and f_2 are differentiable on V^1. At which points do f_1 and f_2 belong to the same germ? At which points do they have the same differential?

5. Let $(x_1, \ldots, x_n)$ be coordinates in V^n. Let $\gamma \in \mathfrak{C}$. Let $(g_1(t), \ldots, g_n(t))$ be the coordinates of $\gamma(t)$. Prove that the tangent vector to γ at v is

$$\sum_{j=1}^{n} g_j'(0) \frac{\partial}{\partial x_j}$$

6. Let $\varphi: U \to V^m$ be differentiable, where U is an open subset of V^n containing the point v. Let $(x_1, \ldots, x_n)$ be coordinates in U and $(y_1, \ldots, y_m)$ be coordinates in V^m. Let φ be given in terms of the coordinates by

$$\varphi: \quad \begin{aligned} y_1 &= h_1(x_1, \ldots, x_n) \\ &\cdots\cdots\cdots\cdots \\ y_m &= h_m(x_1, \ldots, x_n) \end{aligned}$$

Prove that

$$\varphi^*(dy_j) = \sum_{i=1}^{n} \left.\frac{\partial y_j}{\partial x_i}\right|_v dx_i$$

$$\varphi_*\left(\frac{\partial}{\partial x_j}\right) = \sum_{i=1}^{m} \left.\frac{\partial y_i}{\partial x_j}\right|_v \frac{\partial}{\partial y_i}$$

7. Let $\varphi: V^2 \to V^2$ be given by

$$\varphi: \quad \begin{aligned} y_1 &= x_1 e^{x_2} + x_2 \\ y_2 &= x_1 e^{x_2} - x_2 \end{aligned}$$

Prove that φ is a diffeomorphism. Find the matrix of φ_* with respect to the basis $\{\partial/\partial x_1, \partial/\partial x_2\}$ and of φ^* with respect to the basis $\{dx_1, dx_2\}$ at (0, 0).

8. Repeat Exercise 7 for the mapping $\varphi: V^3 \to V^3$ given by

$$\varphi: \quad \begin{aligned} y_1 &= \frac{x_1}{2 + x_2^2} + x_2 e^{x_3} \\ y_2 &= \frac{x_1}{2 + x_2^2} - x_2 e^{x_3} + x_3 \\ y_3 &= 2x_2 e^{x_3} + x_3 \end{aligned}$$

at the point $(x_1, x_2, x_3) = (1, 0, 0)$.

9. Let U be an open set in V^n and S an open set in V^m. Let $\varphi: U \to S$, $\psi: S \to V^l$ be differentiable mappings. Let $\theta = \psi \circ \varphi$. Prove that for the induced linear transformations

$$\theta_* = \psi_* \circ \varphi_* \qquad \theta^* = \varphi^* \circ \psi^*$$

10. Prove that, if $\varphi: V^n \to V^n$ is a diffeomorphism, the induced linear transformations φ_* and φ^* are nonsingular.

11. Let $f, g \in \mathfrak{F}$. Prove that at v

$$d(fg) = f(v)\, dg + g(v)\, df$$

12. Let $f \in \mathfrak{F}$. Let $a = f(v)$. Let ϵ be a positive number, and let $q(t)$ be a real-valued function defined and differentiable for $|t - a| < \epsilon$. Let $g = q \circ f$. Prove that $dg = q'(a)\, df$ at v.

1-4 INNER PRODUCTS AND QUADRATIC FORMS

Let $\{v^1, \ldots, v^n\}$ be a basis of V^n. Define the linear functional $\lambda_i: V^n \to$ real numbers by

$$\lambda_i(c_1 v^1 + \cdots + c_n v^n) = c_i$$

Then $\{\lambda_1, \ldots, \lambda_n\}$ is the basis of the dual space V^{n*} which is dual to $\{v^1, \ldots, v^n\}$. We shall write $\langle u, \lambda_i \rangle$ for $\lambda_i(u)$.

Define the homeomorphism $\sigma_x: V^n \to E^n$ by

$$\sigma_x(u) = (\langle u, \lambda_1 \rangle + x_1, \ldots, \langle u, \lambda_n \rangle + x_n)$$

Then σ_x attaches V^n affinely to the point x, and we may use it to identify E^n with V^n. In this way we introduce a metric into V^n. We may assign a length $|u|$ to each vector u, namely, the distance from $\sigma_x(0)$ to $\sigma_x(u)$. Then

$$|u|^2 = \sum_{i=1}^{n} \langle u, \lambda_i \rangle^2$$

Let $\tau_x\colon V^n \to E^n$ be any homeomorphism that attaches V^n affinely to the point x. We know that there is a nonsingular linear transformation $\mu\colon V^n \to V^n$ such that $\tau_x = \sigma_x \circ \mu$. Hence

$$\tau_x(u) = (\langle \mu(u), \lambda_1 \rangle + x_1, \ . \ . \ . \ , \langle \mu(u), \lambda_n \rangle + x_n)$$

By means of the homeomorphism τ_x we introduce a different metric into V^n with respect to which the length $|u|$ of the vector u is given by

$$|u|^2 = \sum_{i=1}^{n} \langle \mu(u), \lambda_i \rangle^2$$

Let (a_{ij}) be the matrix of the linear transformation μ, that is,

$$\mu(v^j) = \sum_{i=1}^{n} a_{ij} v^i$$

for $j = 1, \ . \ . \ . \ , n$. Then

$$\langle \mu(u), \lambda_i \rangle = \sum_{j=1}^{n} a_{ij} \langle u, \lambda_j \rangle$$

as we may see by first checking it for $u = v^k$. Hence the length $|u|$ of a vector u with respect to the metric introduced into V^n by τ_x is given by

$$|u|^2 = \sum_{i=1}^{n} \sum_{j=1}^{n} g_{ij} \langle u, \lambda_i \rangle \langle u, \lambda_j \rangle$$

where $(g_{ij}) = (a_{ij})^t(a_{ij})$ and $(a_{ij})^t$ denotes the transpose of (a_{ij}).

The matrix (g_{ij}) is symmetric; that is, it satisfies the relation $(g_{ij})^t = (g_{ij})$. It is also positive-definite, which means that the quadratic form

$$\sum_{i=1}^{n} \sum_{j=1}^{n} g_{ij} X_i X_j$$

is nonnegative for all real values of the variables $X_1, \ . \ . \ . \ , X_n$ and is equal to 0 only if all these variables take the value 0. If we define a mapping $(u, v) \to \langle u, v \rangle$ of $V^n \times V^n$ into the real numbers by

$$\langle u, v \rangle = \sum_{i=1}^{n} \sum_{j=1}^{n} g_{ij} \langle u, \lambda_i \rangle \langle v, \lambda_j \rangle$$

these properties are equivalent to the properties of symmetry,

$$\langle v, u \rangle = \langle u, v \rangle$$

and positive definiteness,

$$\langle u, u \rangle \geqq 0 \qquad \text{and} \qquad \langle u, u \rangle = 0 \qquad \text{only if } u = 0$$

for this mapping. The "product" $\langle u, v \rangle$ is a product in the algebraic sense, in that it satisfies

$$\langle a_1 u_1 + a_2 u_2, v \rangle = a_1 \langle u_1, v \rangle + a_2 \langle u_2, v \rangle$$

where a_1 and a_2 are numbers, and a similar linear relation for the second argument. In terms of this product, $|u|^2 = \langle u, u \rangle$.

A "product" $\langle u, v \rangle$, defined on $V^n \times V^n$ with real values, which is symmetric, positive-definite, and bilinear (linear in each of its "factors"), is called an *inner product* on V^n. Every such inner product has the form

$$\langle u, v \rangle = \sum_{i=1}^{n} \sum_{j=1}^{n} g_{ij} \langle u, \lambda_i \rangle \langle u, \lambda_j \rangle$$

where (g_{ij}) is a positive-definite symmetric matrix, as the reader may easily show, since $g_{ij} = \langle v^i, v^j \rangle$. We may summarize in these terms what we have learned so far.

Theorem 1-6. *Let $|u|$ be the length of the vector u determined by identifying E^n with V^n by means of a suitable homeomorphism. Then there is an inner product $\langle u, v \rangle$ on V^n with respect to which $|u|^2 = \langle u, u \rangle$ for all $u \in V^n$.*

We shall show that the converse to this theorem is also true.

Let $\langle u, v \rangle$ be an inner product on V^n. Two vectors u, v are said to be *orthogonal* if $\langle u, v \rangle = 0$. If the vectors $v^1, \ldots, v^m$ are mutually orthogonal, they are linearly independent. If they are mutually orthogonal and also of unit length, that is, if

$$\langle v^i, v^j \rangle = \begin{cases} 0 & \text{when } i \neq j \\ 1 & \text{when } i = j \end{cases}$$

they are said to be an *orthonormal* set.

Lemma 1-3. *Let $\{v^1, \ldots, v^m\}$ be a linearly independent set of vectors from V^n. With respect to each inner product on V^n there is an orthonormal set $\{u^1, \ldots, u^m\}$ which spans the same subspace as $\{v^1, \ldots, v^m\}$.*

Proof. If $m = 1$, we have only to define

$$u^1 = \frac{1}{\sqrt{\langle v^1, v^1\rangle}} v^1$$

Assume that the theorem has been proved for $m = 1, \ldots, r$. Let $\{v^1, \ldots, v^{r+1}\}$ be a linearly independent set. There is an orthonormal set $\{u^1, \ldots, u^r\}$ which spans the same subspace as $\{v^1, \ldots, v^r\}$. Thus $\{u^1, \ldots, u^r, v^{r+1}\}$ is a linearly independent set. Let

$$w = v^{r+1} - \langle v^{r+1}, u^1\rangle u^1 - \cdots - \langle v^{r+1}, u^r\rangle u^r$$

Then $w \neq 0$, and we may define

$$u^{r+1} = \frac{1}{\sqrt{\langle w, w\rangle}} w$$

Then $\{u^1, \ldots, u^{r+1}\}$ is an orthonormal set which spans the same subspace as $\{v^1, \ldots, v^{r+1}\}$. The theorem is therefore true for all m by induction.

Theorem 1-7. *Let $\langle u, v\rangle$ be an inner product on V^n. Then there is a homeomorphism $\tau_x: V^n \to E^n$ which attaches V^n affinely to the point x such that the length $|u|$ which is determined by τ_x satisfies $|u|^2 = \langle u, u\rangle$ for all $u \in V^n$.*

Proof. Let $\{v^1, \ldots, v^n\}$ be a basis of V^n. By Lemma 1-3 we may assume that $\{v^1, \ldots, v^n\}$ is an orthonormal set. Let $\{\lambda_1, \ldots, \lambda_n\}$ be the dual basis to $\{v^1, \ldots, v^n\}$. Define $\tau_x: V^n \to E^n$ by

$$\tau_x(u) = (\langle u, \lambda_1\rangle + x_1, \ldots, \langle u, \lambda_n\rangle + x_n)$$

Then
$$|u|^2 = \sum_{i=1}^n \langle u, \lambda_i\rangle^2$$

But we also have

$$u = \sum_{i=1}^n \langle u, \lambda_i\rangle v^i$$

from which we obtain

$$\langle u, u\rangle = \sum_{i=1}^n \langle u, \lambda_i\rangle^2$$

Hence
$$|u|^2 = \langle u, u\rangle$$

EXERCISES

1. Prove that, if (g_{ij}) is a positive-definite symmetric matrix, there is a nonsingular matrix (a_{ij}) such that $(g_{ij}) = (a_{ij})^t(a_{ij})$.

2. Let $u \to |u|$ be a mapping of V^n into the nonnegative real numbers such that $|u| = 0$ only if $u = 0$. Prove that there is an inner product $\langle u, v\rangle$ on V^n such that $\langle u, u\rangle = |u|^2$ if and only if the following conditions are satisfied:

a. $|au| = |a|\,|u|$ for numbers a.

b. $|u + v| \leqq |u| + |v|$.

c. $|u + v|^2 + |u - v|^2 = 2|u|^2 + 2|v|^2$.

1-5 THE STRUCTURE OF THE GROUP OF AFFINITIES OF A^n

Let $\sigma_x\colon V^n \to A^n$ be a homeomorphism which attaches V^n affinely to the point x in A^n. If $\varphi\colon A^n \to A^n$ is an affinity, let $u = \sigma_x^{-1}(\varphi(x))$. Let $\rho_{-u}\colon A^n \to A^n$ be the action of the element $-u$ on A^n. Then $\psi = \rho_{-u}\circ\varphi$ is an affinity of A^n which keeps the point x fixed. If we write $\varphi = \rho_u\circ\psi$, we see that we have proved that *every affinity is the composition of an affinity which keeps the point x fixed and a translation.* We shall therefore study the group of affinities that keep x fixed.

If $\varphi\colon A^n \to A^n$ is an affinity that keeps x fixed, then $\varphi_* = \sigma_x^{-1}\circ\varphi\circ\sigma_x$ is a nonsingular linear transformation of V^n, by Theorem 1-3. Conversely, if $\varphi_*\colon V^n \to V^n$ is a nonsingular linear transformation,

$$\varphi = \sigma_x\circ\varphi_*\circ\sigma_x^{-1}$$

is an affinity, as the reader may easily demonstrate. Furthermore, if φ_1 and φ_2 are two affinities of A^n that keep x fixed, the induced linear transformation $(\varphi_1\circ\varphi_2)_*$ is the same as the composition $\varphi_{1*}\circ\varphi_{2*}$ of the linear transformations φ_{1*} and φ_{2*}. Consequently the group of affinities that keep a point fixed has the same structure as the group $GL(n)$ of all nonsingular linear transformations of V^n. In group theoretic terms, $A(n) = V(n)GL(n)$, where $A(n)$ is the group of all affinities of A^n and $V(n)$ is the group of translations.

Theorem 1-8. *$V(n)$ is a normal subgroup of $A(n)$. If ρ_u denotes the action of $u \in V^n$ on A^n and φ and ψ are affinities that keep the point x fixed, then*

$$(\rho_{u_1}\circ\varphi)\circ(\rho_{u_2}\circ\psi) = \rho_v\circ(\varphi\circ\psi)$$

where $v = u_1 + \varphi_*(u_2)$ and φ_* is the linear transformation of V^n induced by φ.

Proof. Let $\sigma_x: V^n \to A^n$ attach V^n affinely to the point x. Let $\varphi: A^n \to A^n$ be an affinity that keeps the point x fixed. Then $\varphi_* = \sigma_x^{-1} \circ \varphi \circ \sigma_x$. By simple computations we may prove that

$$\varphi \circ \rho_v \circ \varphi^{-1} = \rho_{\varphi_*(v)}$$

for $v \in V^n$. If ρ_u is any translation, then

$$(\rho_u \circ \varphi) \circ \rho_v \circ (\rho_u \circ \varphi)^{-1} = \rho_{\varphi_*(v)}$$

which proves that $V(n)$ is a normal subgroup of $A(n)$. The multiplication rule stated in Theorem 1-8 also follows from these computations.

We may derive from Theorem 1-8 a representation of the group $A(n)$ by matrices. Let $\{v^1, \ldots, v^n\}$ be a basis of V^n. For each element φ_* of $GL(n)$ there is a unique nonsingular $n \times n$ matrix (c_{ij}) such that

$$\varphi_*(v^j) = \sum_{i=1}^{n} c_{ij} v^i$$

for $j = 1, \ldots, n$. If φ'_* is a second element of $GL(n)$ with the corresponding matrix (c'_{ij}), then the matrix of $\varphi_* \circ \varphi'_*$ is $(c_{ij})(c'_{ij})$.

On the other hand, if $u \in V^n$, there are unique numbers $a_1, \ldots, a_n$ such that

$$u = a_1 v^1 + \cdots + a_n v^n$$

We shall map the affinity $\rho_u \circ \varphi$ onto the matrix

$$\begin{pmatrix} & & a_1 \\ & & \cdot \\ & (c_{ij}) & \cdot \\ & & \cdot \\ & & a_n \\ 0 & \cdots \quad 0 & 1 \end{pmatrix}$$

Thus each element of $A(n)$ corresponds to an $(n+1) \times (n+1)$ matrix, and this correspondence is one-to-one. If the matrix

$$\begin{pmatrix} & & a'_1 \\ & & \cdot \\ & (c'_{ij}) & \cdot \\ & & \cdot \\ & & a'_n \\ 0 & \cdots \quad 0 & 1 \end{pmatrix}$$

corresponds to the affinity $\rho_{u'}\circ\varphi'$, we hope to find that the matrix

$$\begin{pmatrix} & (c_{ij}) & & a_1 \\ & & & \cdot \\ & & & \cdot \\ & & & \cdot \\ & & & a_n \\ 0 & \cdots & 0 & 1 \end{pmatrix}\begin{pmatrix} & (c'_{ij}) & & a'_1 \\ & & & \cdot \\ & & & \cdot \\ & & & \cdot \\ & & & a'_n \\ 0 & \cdots & 0 & 1 \end{pmatrix} = \begin{pmatrix} & (c_{ij})(c'_{ij}) & & a_1 + b_1 \\ & & & \cdot \\ & & & \cdot \\ & & & \cdot \\ & & & a_n + b_n \\ 0 & \cdots & 0 & 1 \end{pmatrix}$$

where

$$\begin{pmatrix} b_1 \\ \cdot \\ \cdot \\ \cdot \\ b_n \end{pmatrix} = (c_{ij})\begin{pmatrix} a'_1 \\ \cdot \\ \cdot \\ \cdot \\ a'_n \end{pmatrix}$$

corresponds to $(\rho_u\circ\varphi)\circ(\rho_{u'}\circ\varphi')$. This follows from Theorem 1-8.

If we identify E^n and A^n, we may regard the group $R(n)$ of all rigid motions as a subgroup of $A(n)$. The translations of A^n are rigid motions, and so it follows that *every rigid motion is the composition of a rigid motion with a fixed point and a translation.* A rigid motion with a fixed point is called a *rotation.* The group $O(n)$ of all rotations of E^n is called the *orthogonal group.* $R(n) = V(n)O(n)$ in group theoretic terms. The relationship among the various groups with which we have been concerned is indicated in Figure 1-5.

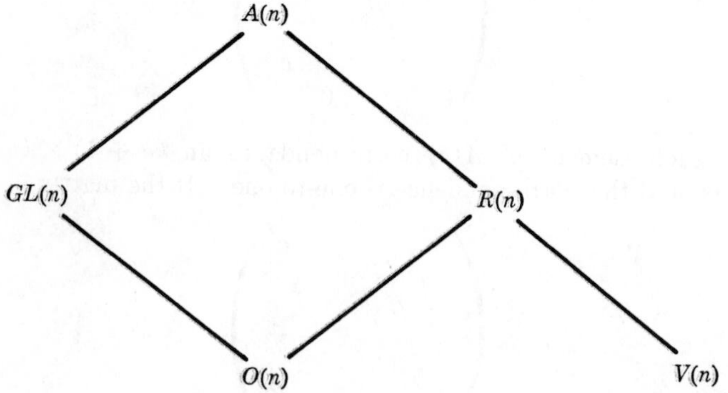

FIG. 1-5

Let us choose as our basis $\{v^1, \ldots, v^n\}$ of V^n an orthonormal set with respect to the inner product on V^n which is derived from the identification of E^n with V^n. Since this inner product is given in terms of the lengths of vectors by the relation

$$2\langle u, v\rangle = |u + v|^2 - |u|^2 - |v|^2$$

it follows that a rotation $\rho: E^n \to E^n$ about the point x to which V^n is attached preserves the inner product, that is,

$$\langle \rho_*(u), \rho_*(v)\rangle = \langle u, v\rangle$$

where $\rho_*: V^n \to V^n$ is the linear transformation induced by ρ. It follows that $\{\rho_*(v^1), \ldots, \rho_*(v^n)\}$ is also an orthonormal set. If (c_{ij}) is the matrix of ρ_* with respect to $\{v^1, \ldots, v^n\}$, the relations

$$\langle \rho_*(v^i), \rho_*(v^j)\rangle = \begin{cases} 0 & \text{if } i \neq j \\ 1 & \text{if } i = j \end{cases}$$

tell us that $(c_{ij})^t(c_{ij})$ is the identity matrix. Such matrices are said to be *orthogonal*. Conversely, if (c_{ij}) is an orthogonal matrix, the nonsingular linear transformation ρ_* which it represents with respect to the basis $\{v^1, \ldots, v^n\}$ preserves the inner product on V^n. Thus it is the linear transformation induced by a rotation about the point x. It follows that the rigid motions of E^n may be represented by matrices of the form

$$\begin{pmatrix} & & & a_1 \\ & & & \cdot \\ & (c_{ij}) & & \cdot \\ & & & \cdot \\ & & & a_n \\ 0 & \cdots & 0 & 1 \end{pmatrix}$$

where (c_{ij}) is an orthogonal matrix.

EXERCISE

Let $\langle u, v\rangle$ be an inner product on V^n. Prove that a mapping $\theta: V^n \to V^n$, for which $\langle \theta(u), \theta(v)\rangle = \langle u, v\rangle$ for all $u, v \in V^n$, is a nonsingular linear transformation.

CHAPTER 2

Differentiable Manifolds

In this chapter we shall see how it is possible to assign a differentiable structure to certain topological spaces which are of a more general character than euclidean n-space. What we shall demand is that these spaces be locally homeomorphic to V^n and that these local homeomorphisms fit together in a smooth way. For the purpose of visualizing such spaces intuitively, the reader should keep in mind the familiar example of a surface in ordinary euclidean 3-space without singularities, defined by a differentiable relation $f(x) = 0$.

We shall first consider the general class of examples consisting of subspaces of E^n which are defined by means of a set of relations. Later we shall give the definition of an abstract differentiable manifold which does not require that the manifold be a subset of some euclidean space. Although we shall not need it for the general theory, it is interesting to examine the relationship that exists between the differentiable structure of a subspace of E^n and the differentiable structure of E^n itself.

2-1 DIFFERENTIABLE VARIETIES IN E^n

Let $f_1, \ldots, f_s$ be functions which are differentiable on an open set V in E^n. These functions define a *variety* S of E^n consisting of those points x in V for which $f_1(x) = \cdots = f_s(x) = 0$. We shall assign to S the topology which is induced on it by the topology of E^n. For example, a single function $f(x_1, x_2, x_3)$ defines a surface in E^3. A pair of functions f_1, f_2 defines the intersection of two surfaces, which is generally a curve.

It is necessary to impose a condition on S which will guarantee that there are no singularities. We shall assume that each point of S has a neighborhood in E^n on which the matrix

$$\begin{pmatrix} \dfrac{\partial f_1}{\partial x_1} & \cdots & \dfrac{\partial f_1}{\partial x_n} \\ \cdots & \cdots & \cdots \\ \dfrac{\partial f_s}{\partial x_1} & \cdots & \dfrac{\partial f_s}{\partial x_n} \end{pmatrix}$$

has constant rank r with $r < n$ and that r is the same for all points of S. A variety S that is subject to this condition will be called a *differentiable* variety in E^n. In the next section we shall show that, if x is a point of S, there are an open subset W of E^m, where $m = n - r$, and a one-to-one differentiable mapping $\eta: W \to E^n$ such that $U = \eta(W)$ is an open subset of S which contains x. Furthermore, there are an open subset $\mathcal{U}$ of E^n and a differentiable mapping $\zeta: \mathcal{U} \to E^m$ such that $U \subset \mathcal{U}$ and $\zeta \circ \eta$ is the identity mapping on W. It follows that η is one-to-one and

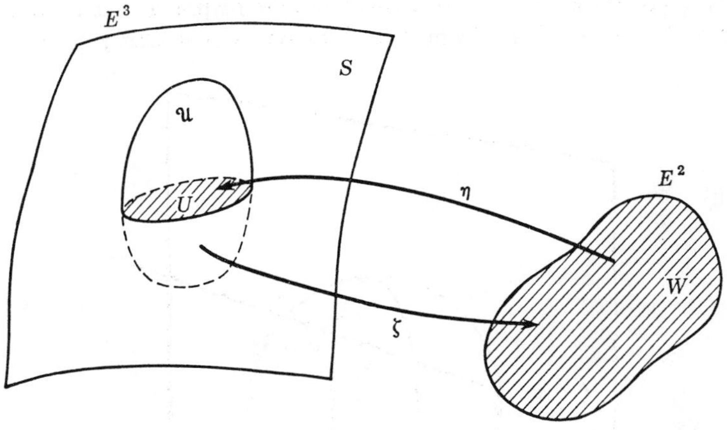

FIG. 2-1

its inverse mapping is the restriction of ζ to U. In the case of a surface S in E^3, this says that there is a one-to-one differentiable mapping of an open set W in the plane E^2 onto an open set U on the surface which can be inverted by a differentiable mapping from E^3 into E^2 (see Figure 2-1).

Actually, there are many choices for the set W and the mapping η. Each set $U = \eta(W)$, together with its corresponding mapping η, will be called a *coordinate neighborhood* of S. It should be apparent that we may attach to S at the point x in U a tangent space of dimension m which plays the same role as the tangent space to E^m at a point. This construction will be presented in detail for abstract differentiable manifolds, and so we shall postpone it until that time.

Let us consider, as an example, the 2-sphere S^2 defined by $x_1^2 + x_2^2 + x_3^2 = 1$. In this case the matrix under consideration is the 1×3 matrix $(2x_1 \quad 2x_2 \quad 2x_3)$ which has rank 1 at all points of S^2. We shall show how to construct a coordinate neighborhood of the point $(1, 0, 0)$. Let W be the plane $x_1 = 0$. If $u = (0, u_1, u_2)$ is a point on W, we let $\eta(u)$ be the point where the line joining u to $(-1, 0, 0)$ intersects S^2 (see Figure 2-2). If $\eta(u) = (x_1, x_2, x_3)$, then

$$x_1 = \frac{1 - u_1^2 - u_2^2}{1 + u_1^2 + u_2^2}$$

$$x_2 = \frac{2u_1}{1 + u_1^2 + u_2^2}$$

$$x_3 = \frac{2u_2}{1 + u_1^2 + u_2^2}$$

Consequently $\eta\colon W \to E^3$ is a differentiable mapping that carries W onto the subset U of S^2 which is obtained by removing just the one

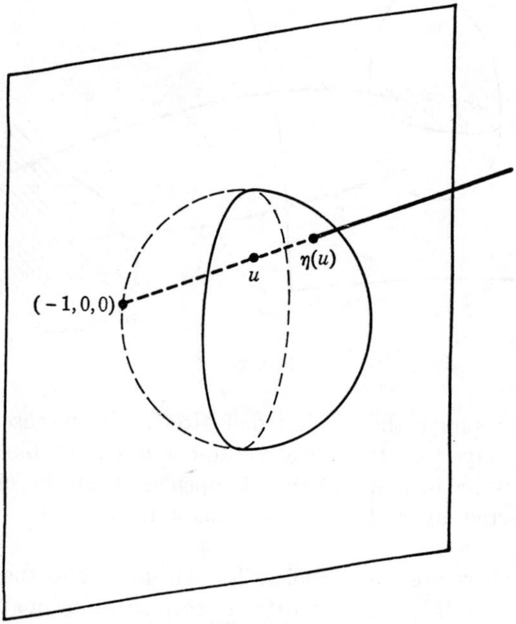

FIG. 2-2

point $(-1, 0, 0)$. On the other hand, let $\mathfrak{U}$ be the subset of E^3 obtained by removing the point $(-1, 0, 0)$ and define $\zeta\colon \mathfrak{U} \to W$ by projection through the point $(-1, 0, 0)$. If $\zeta(x_1, x_2, x_3) = (0, u_1, u_2)$, then

$$u_1 = \frac{x_2}{1 + x_1} \qquad u_2 = \frac{x_3}{1 + x_1}$$

and so ζ is a differentiable mapping. The projection η is called the stereographic projection of the plane onto the sphere.

Before we leave the differentiable variety S in E^n and turn to abstract manifolds, we shall examine the relationship between the notions of differentiability on S and differentiability in E^n. Let g be a function which is defined on a neighborhood in S of a point x. Let U be a coordinate neighborhood of x, and let $\eta: W \to U$ be the corresponding mapping. We shall say that g is *differentiable* at x if the composite mapping $g \circ \eta: W \to R$ is differentiable at $\eta^{-1}(x)$. Since x may lie in many coordinate neighborhoods, we must show that this notion of differentiability does not depend upon the particular coordinate neighborhood that we use.

Let U_1 and U_2 be two coordinate neighborhoods of x, and let $\eta_1: W_1 \to U_1$ and $\eta_2: W_2 \to U_2$ be the corresponding mappings. There are open sets $\mathfrak{U}_1$ and $\mathfrak{U}_2$ in E^n and differentiable mappings $\zeta_1: \mathfrak{U}_1 \to W_1$ and $\zeta_2: \mathfrak{U}_2 \to W_2$ such that $U_1 \subset \mathfrak{U}_1$, $U_2 \subset \mathfrak{U}_2$, $\zeta_1 \circ \eta_1$ is the identity mapping on W_1, and $\zeta_2 \circ \eta_2$ is the identity mapping on W_2. If we suppose that $g \circ \eta_1$ is differentiable at $\eta_1^{-1}(x)$, then, on $\eta_2^{-1}(V_1 \cap V_2)$, $g \circ \eta_2 = g \circ \eta_1 \circ \zeta_1 \circ \eta_2$ is the composition of the differentiable mappings η_2 and ζ_1 and the differentiable function $g \circ \eta_1$, and so $g \circ \eta_2$ is differentiable at $\eta_2^{-1}(x)$. Similarly, if $g \circ \eta_2$ is differentiable at $\eta_2^{-1}(x)$, then $g \circ \eta_1$ is differentiable at $\eta_1^{-1}(x)$.

The next theorem shows that the differentiable functions on S are locally the restrictions to S of differentiable functions on E^n.

Theorem 2-1. *Let g be a function which is defined on the coordinate neighborhood U of S. Then g is differentiable on U if and only if g is the restriction to U of a function h which is differentiable on an open subset $\mathfrak{U}$ of E^n such that $U \subset \mathfrak{U}$.*

Proof. Let $\eta: W \to U$ be the differentiable mapping associated with the coordinate neighborhood U. Let $\zeta: \mathfrak{U} \to W$ be the differentiable mapping such that $U \subset \mathfrak{U}$, where $\mathfrak{U}$ is an open subset of E^n, and $\zeta \circ \eta$ is the identity mapping on W.

1. Assume that g is differentiable on U. Define $h = g \circ \eta \circ \zeta$ on $\mathfrak{U}$. Then h is differentiable on $\mathfrak{U}$, and $h(x) = g(x)$ for x in U.

2. Conversely, assume that g is the restriction of a function h which is differentiable on an open subset $\mathfrak{V}$ of E^n such that $U \subset \mathfrak{V}$. Now $g \circ \eta = h \circ \eta$, and $h \circ \eta$ is differentiable on W. Hence g is differentiable on U.

Let V be an open subset of E^l. The differentiability of a mapping $\varphi: V \to S$ will be defined by referring it to the differentiability of functions on V. We shall say that φ is *differentiable* at the point y in V if for every function g on S which is differentiable at $\varphi(y)$ the function $g \circ \varphi$ is differentiable at y.

Now $\varphi: V \to S$ may also be regarded as a mapping into E^n. From the point of view $\varphi: V \to E^n$, we know that φ is differentiable at y if for every function h on E^n which is differentiable at $\varphi(y)$ the function $h\circ\varphi$ is differentiable at y. According to Theorem 2-1, the functions g used in the definition of the differentiability of $\varphi: V \to S$ are the restrictions to S of the functions h which are used in the definition of the differentiability of $\varphi: V \to E^n$. Consequently these two notions of differentiability for φ are equivalent to one another.

EXERCISES

1. Show that the relation

$$x_1^3 + x_2^3 + x_3^3 - 3x_1x_2x_3 = 1$$

defines a differentiable variety in E^3.

2. Show that the relations

$$\begin{aligned} x_1 + x_2 + x_3 &= 0 \\ x_1x_2x_3 &= 2 \end{aligned}$$

define a differentiable variety in E^3 provided the points $(-1, -1, 2)$, $(-1, 2, -1)$, and $(2, -1, -1)$ are excluded.

3. Show that the relations

$$\begin{aligned} x_1^2 + 2x_1x_3 - 2x_2x_3 - x_2^2 &= 0 \\ 2x_1 - x_2 + x_3 &= 3 \end{aligned}$$

define a differentiable variety in E^3.

4. Let W be the subset of the plane $x_1 = 0$ in E^3 consisting of those points $(0, u_1, u_2)$ for which $u_1^2 + u_2^2 < 1$. If $u = (0, u_1, u_2)$ is a point of W, let $\eta(u)$ be the point where the line through u parallel to the x_1-axis intersects S^2 with a positive x_1-coordinate. Prove that $\eta(W)$ is a coordinate neighborhood of S^2 with respect to the mapping η.

2-2 THE IMPLICIT-FUNCTION THEOREMS

The unproved statement in the previous section concerning the existence of coordinate neighborhoods on varieties in E^n is a consequence of the well-known implicit-function theorem of advanced calculus. The implicit-function theorem describes sufficient conditions under which a system of r equations

$$\begin{aligned} f_1(x_1, \ldots, x_r, y_1, \ldots, y_m) &= 0 \\ \cdots\cdots\cdots\cdots\cdots\cdots\cdots& \\ f_r(x_1, \ldots, x_r, y_1, \ldots, y_m) &= 0 \end{aligned} \tag{2-2-1}$$

in $r + m$ variables $x_1, \ldots, x_r, y_1, \ldots, y_m$ may be solved for $x_1, \ldots, x_r$ in terms of $y_1, \ldots, y_m$.

Implicit-function Theorem of Advanced Calculus. *If $f_1, \ldots, f_r$ are differentiable functions on a neighborhood of the point*

$$(x^\circ, y^\circ) = (x_1^\circ, \ldots, x_r^\circ, y_1^\circ, \ldots, y_m^\circ)$$

in E^{r+m}, if $f_1(x^\circ, y^\circ) = \cdots = f_r(x^\circ, y^\circ) = 0$, and if the $r \times r$ matrix

$$\begin{pmatrix} \dfrac{\partial f_1}{\partial x_1} & \cdots & \dfrac{\partial f_1}{\partial x_r} \\ \cdots & \cdots & \cdots \\ \dfrac{\partial f_r}{\partial x_1} & \cdots & \dfrac{\partial f_r}{\partial x_r} \end{pmatrix}$$

is nonsingular at (x°, y°), then there are a neighborhood U of the point $y^\circ = (y_1^\circ, \ldots, y_m^\circ)$ in E^m, a neighborhood V of the point

$$x^\circ = (x_1^\circ, \ldots, x_r^\circ)$$

in E^r, and a unique mapping $\varphi: U \to V$ such that $\varphi(y^\circ) = x^\circ$ and $f_1(\varphi(y), y) = \cdots = f_r(\varphi(y), y) = 0$ for all y in U. Furthermore, φ is differentiable.

In other words, if we write

$$\varphi(y) = (g_1(y), \ldots, g_r(y))$$

where $g_1, \ldots, g_r$ are differentiable functions on U, then

$$\begin{aligned} x_1 &= g_1(y_1, \ldots, y_m) \\ &\cdots\cdots\cdots \\ x_r &= g_r(y_1, \ldots, y_m) \end{aligned}$$

is the unique solution of the system (2-2-1) near y°. A proof of this theorem may be found in most books on advanced calculus. In particular, the reader may consult R. C. Buck, "Advanced Calculus" (McGraw-Hill, New York, 1956).

We shall extend this theorem to take care of the case where the number of equations exceeds the number of unknowns.

General Implicit-function Theorem. *If $f_1, \ldots, f_s$ are differentiable functions on a neighborhood of the point*

$$(x^\circ, y^\circ) = (x_1^\circ, \ldots, x_r^\circ, y_1^\circ, \ldots, y_m^\circ)$$

in E^{r+m}, *if* $f_1(x^\circ, y^\circ) = \cdots = f_s(x^\circ, y^\circ) = 0$, *if the* $s \times r$ *matrix*

$$\mathcal{J}(x, y) = \begin{pmatrix} \dfrac{\partial f_1}{\partial x_1} & \cdots & \dfrac{\partial f_1}{\partial x_r} \\ \cdots & \cdots & \cdots \\ \dfrac{\partial f_s}{\partial x_1} & \cdots & \dfrac{\partial f_s}{\partial x_r} \end{pmatrix}$$

has rank r *at* (x°, y°), *and if the* $s \times (r + m)$ *matrix*

$$\mathcal{J}^*(x, y) = \begin{pmatrix} \dfrac{\partial f_1}{\partial x_1} & \cdots & \dfrac{\partial f_1}{\partial x_r} & \dfrac{\partial f_1}{\partial y_1} & \cdots & \dfrac{\partial f_1}{\partial y_m} \\ \cdots & \cdots & \cdots & \cdots & \cdots & \cdots \\ \dfrac{\partial f_s}{\partial x_1} & \cdots & \dfrac{\partial f_s}{\partial x_r} & \dfrac{\partial f_s}{\partial y_1} & \cdots & \dfrac{\partial f_s}{\partial y_m} \end{pmatrix}$$

has constant rank r *on a neighborhood of* (x°, y°), *then there are a neighborhood* U *of the point* $y^\circ = (y_1{}^\circ, \ldots, y_m{}^\circ)$ *in* E^m, *a neighborhood* V *of the point* $x^\circ = (x_1{}^\circ, \ldots, x_r{}^\circ)$ *in* E^r, *and a unique mapping* $\varphi\colon U \to V$ *such that* $\varphi(y^\circ) = x^\circ$ *and* $f_1(\varphi(y), y) = \cdots = f_s(\varphi(y), y) = 0$ *for all* y *in* U. *Furthermore,* φ *is differentiable.*

Proof. By renumbering the functions $f_1, \ldots, f_s$, we may assume that the submatrix

$$\mathcal{J}_1(x, y) = \begin{pmatrix} \dfrac{\partial f_1}{\partial x_1} & \cdots & \dfrac{\partial f_1}{\partial x_r} \\ \cdots & \cdots & \cdots \\ \dfrac{\partial f_r}{\partial x_1} & \cdots & \dfrac{\partial f_r}{\partial x_r} \end{pmatrix}$$

consisting of the first r rows of $\mathcal{J}(x, y)$ is nonsingular at (x°, y°). It follows from the continuity of the determinant of $\mathcal{J}_1(x, y)$ that there is a neighborhood W of (x°, y°) on which $\mathcal{J}_1(x, y)$ is nonsingular. We may assume that the neighborhood W is so small that $\mathcal{J}^*(x, y)$ has rank r at every point of W. It follows that the first r rows of $\mathcal{J}^*(x, y)$ are linearly independent and its remaining rows are linear combinations of these when (x, y) is in W. We may write for the ith row

$$\begin{aligned} \frac{\partial f_i}{\partial x_j} &= a_1 \frac{\partial f_1}{\partial x_j} + \cdots + a_r \frac{\partial f_r}{\partial x_j} \qquad j = 1, \ldots, r \\ \frac{\partial f_i}{\partial y_k} &= a_1 \frac{\partial f_1}{\partial y_k} + \cdots + a_r \frac{\partial f_r}{\partial y_k} \qquad k = 1, \ldots, m \end{aligned} \tag{2-2-2}$$

for (x, y) in W, where $a_1, \ldots, a_r$ are functions on W.

Apply the implicit-function theorem of advanced calculus to the functions $f_1, \ldots, f_r$. There are a neighborhood U of y°, a neighborhood V of x°, and a unique mapping $\varphi: U \to V$ such that $\varphi(y^\circ) = x^\circ$ and

$$f_1(\varphi(y), y) = \cdots = f_r(\varphi(y), y) = 0 \tag{2-2-3}$$

for y in U. φ is differentiable. We may assume that U is a connected set and is so small that $(\varphi(y), y)$ is in W when y is in U. Under these circumstances, we may prove that also

$$f_{r+1}(\varphi(y), y) = \cdots = f_s(\varphi(y), y) = 0$$

for y in U.

Define the functions $g_1, \ldots, g_r$ on U by $\varphi(y) = (g_1(y), \ldots, g_r(y))$ and define $w_i(y) = f_i(\varphi(y), y)$ for $r + 1 \leqq i \leqq s$. Then

$$\frac{\partial w_i}{\partial y_j} = \sum_{k=1}^{r} \frac{\partial f_i}{\partial x_k}\bigg|_{(\varphi(y),y)} \frac{\partial g_k}{\partial y_j} + \frac{\partial f_i}{\partial y_j}\bigg|_{(\varphi(y),y)}$$

If we use the relations (2-2-2), then

$$\frac{\partial w_i}{\partial y_j} = \sum_{l=1}^{r} a_l \left\{ \sum_{k=1}^{r} \frac{\partial f_l}{\partial x_k}\bigg|_{(\varphi(y),y)} \frac{\partial g_k}{\partial y_j} + \frac{\partial f_l}{\partial y_j}\bigg|_{(\varphi(y),y)} \right\}$$

But the expressions in braces are equal to zero for y in U, as can be seen by differentiating the relations (2-2-3) with respect to y_j. Hence $\partial w_i/\partial y_j = 0$ for $j = 1, \ldots, m$. This implies that w_i is constant on U. Since $w_i = 0$ at y°, it follows that $w_i = 0$ on all of U, which was to be proved.

The theorem which is needed for the existence of coordinate neighborhoods on varieties in E^n is a symmetrical form of the general implicit-function theorem. It states sufficient conditions for a system of equations

$$\begin{gathered} f_1(x_1, \ldots, x_n) = 0 \\ \cdots\cdots\cdots\cdots \\ f_s(x_1, \ldots, x_n) = 0 \end{gathered}$$

to have a parametrically described solution

$$\begin{gathered} x_1 = g_1(y_1, \ldots, y_m) \\ \cdots\cdots\cdots\cdots \\ x_n = g_n(y_1, \ldots, y_m) \end{gathered}$$

Implicit-parametrization Theorem. *If $f_1, \ldots, f_s$ are differentiable functions on a neighborhood W of the point $x^\circ = (x_1^\circ, \ldots, x_n^\circ)$ in E^n, if $f_1(x^\circ) = \cdots = f_s(x^\circ) = 0$, and if the $s \times n$ matrix*

$$\mathcal{J}(x) = \begin{pmatrix} \dfrac{\partial f_1}{\partial x_1} & \cdots & \dfrac{\partial f_1}{\partial x_n} \\ \cdots & \cdots & \cdots \\ \dfrac{\partial f_s}{\partial x_1} & \cdots & \dfrac{\partial f_s}{\partial x_n} \end{pmatrix}$$

has the constant rank r on W, with $r < n$, then there are a neighborhood U of $y^\circ = (0, \ldots, 0)$ in E^m, where $m = n - r$, and a differentiable mapping $\eta: U \to W$ such that $\eta(y^\circ) = x^\circ$ and

$$f_1(\eta(y)) = \cdots = f_s(\eta(y)) = 0$$

for y in U. Let S be the variety defined by $f_1, \ldots, f_s$ in W. If W is sufficiently small, then there is a differentiable mapping $\zeta: W \to E^m$ such that $\eta(U) = W \cap S$ and $\zeta \circ \eta$ is the identity mapping on U.

Proof. The proof is similar to the proof of the preceding theorem. Consequently the details will be left to the reader.

By suitably renumbering the coordinates $x_1, \ldots, x_n$, we may apply the general implicit-function theorem in order to solve for $x_1, \ldots, x_r$ in terms of $x_{r+1}, \ldots, x_n$. There are a neighborhood U_1 of the point $(x_{r+1}^\circ, \ldots, x_n^\circ)$ in E^m, a neighborhood V of the point $(x_1^\circ, \ldots, x_r^\circ)$ in E^r, $V \times U_1 \subset W$, and a unique mapping $\eta_1: U_1 \to V$ such that $\eta_1(x_{r+1}^\circ, \ldots, x_n^\circ) = (x_1^\circ, \ldots, x_r^\circ)$ and

$$f_j(\eta_1(x_{r+1}, \ldots, x_n), x_{r+1}, \ldots, x_n) = 0$$

for $j = 1, \ldots, s$ and for $(x_{r+1}, \ldots, x_n) \in U_1$. Let $\tau: E^m \to E^m$ be the translation of E^m which carries $y^\circ = (0, \ldots, 0)$ onto the point $(x_{r+1}^\circ, \ldots, x_n^\circ)$. Let $U = \tau^{-1}(U_1)$, and define $\eta: U \to W$ by

$$\eta(y) = (\eta_1 \circ \tau(y), \tau(y))$$

We may now replace W by the product set $V \times U_1$. Then define $\zeta: V \times U_1 \to U$ by

$$\zeta(x) = \tau^{-1}(x_{r+1}, \ldots, x_n)$$

These sets and mappings satisfy the desired conditions.

2-3 DIFFERENTIABLE MANIFOLDS

With the differentiable varieties in mind, we shall now give the definition of an abstract differentiable manifold.

Let M be a Hausdorff topological space with a denumerable basis. We postulate the existence of an indexed collection of pairs $\{W_\alpha, \eta_\alpha\}_{\alpha \in I}$, where W_α is an open subset of E^n and $\eta_\alpha: W_\alpha \to M$ is a homeomorphism of W_α onto an open subset U_α of M which satisfies the following conditions:

1. If m is a point in M, there is an index α in I such that U_α contains m.

2. For every ordered pair of indices α, β from I such that

$$U = U_\alpha \cap U_\beta$$

is not empty, the restriction of $\eta_\beta^{-1} \circ \eta_\alpha$ to $\eta_\alpha^{-1}(U)$ is a differentiable mapping of this set into E^n.

3. If $\eta: W \to U$ is a homeomorphism of an open subset W of E^n onto an open subset U of M such that for any index α for which $U \cap U_\alpha$ is not empty the restriction of $\eta^{-1} \circ \eta_\alpha$ to $\eta_\alpha^{-1}(U \cap U_\alpha)$ and the restriction of $\eta_\alpha^{-1} \circ \eta$ to $\eta^{-1}(U \cap U_\alpha)$ are differentiable mappings, there is an index β such that $(W, \eta) = (W_\beta, \eta_\beta)$.

The integer n is the same for all indices and is called the *dimension* of the manifold M. The sets U_α are called the *coordinate neighborhoods* of M. The mapping $\varphi_\alpha = \eta_\alpha^{-1}: U_\alpha \to W_\alpha$ will be called the *coordinate mapping* corresponding to the coordinate neighborhood U. Condition 1 guarantees that M is covered by its coordinate neighborhoods. If α, β are indices such that $U_\alpha \cap U_\beta = U$ is not empty, condition 2 tells us that $\eta_\beta^{-1} \circ \eta_\alpha$ and $\eta_\alpha^{-1} \circ \eta_\beta$, when suitably restricted, are both differentiable mappings. Since these two mappings are inverse to one another, it follows that their jacobian matrices have rank n. The mapping $\eta_\beta^{-1} \circ \eta_\alpha = \varphi_\beta \circ \varphi_\alpha^{-1}$ will be denoted by $\psi_\alpha{}^\beta$ and will be called the *coordinate transformation* on $U_\alpha \cap U_\beta$ (see Figure 2-3).

Condition 3 expresses the fact that the family $\{U_\alpha, \varphi_\alpha\}_{\alpha \in I}$ contains every possible coordinate neighborhood of the manifold M. In a certain sense, which we shall make clear, this condition is superfluous. Its purpose is to eliminate the annoying ambiguities which would arise if the family of coordinate neighborhoods were not complete. We shall see that, if we merely assume conditions 1 and 2, it is possible to enlarge the family $\{W_\alpha, \eta_\alpha\}_{\alpha \in I}$ in one and only one way so as to make it satisfy condition 3.

Let M be a Hausdorff space with a denumerable basis and with a family $\mathcal{W} = \{W_\alpha, \eta_\alpha\}_{\alpha \in I}$ subject only to conditions 1 and 2. Under these circumstances, we shall call M a *coordinate manifold.* A pair $\{W, \eta\}$ consisting of an open subset W of E^n and a homeomorphism $\eta: W \to U$ onto an open subset U of M will be said to be *compatible* with $\mathcal{W}$ if, for every index α such that $U \cap U_\alpha$ is not empty, the mappings $\eta^{-1}\circ\eta_\alpha$ and $\eta_\alpha^{-1}\,\eta$ are differentiable when restricted to the sets $\eta_\alpha^{-1}(U \cap U_\alpha)$ and $\eta^{-1}(U \cap U_\alpha)$, respectively. It is clear that, if it is possible to enlarge the family $\mathcal{W}$ so that M becomes a differentiable manifold, the enlarged family consists of those pairs $\{W, \eta\}$ which are compatible with $\mathcal{W}$.

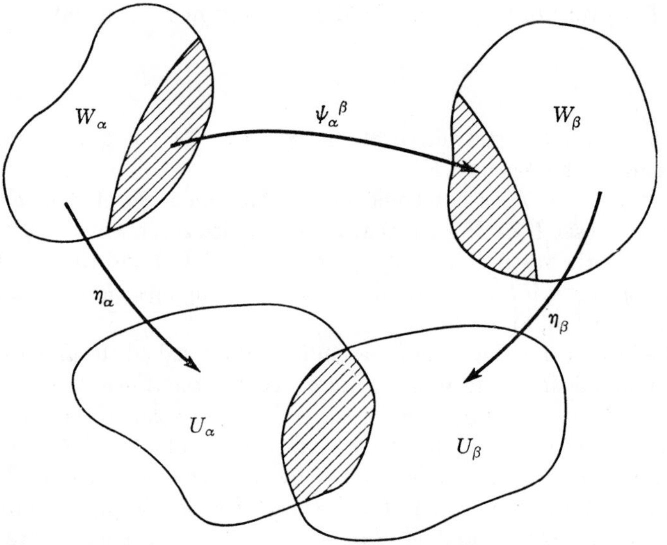

FIG. 2-3

Theorem 2-2. *Let M be a coordinate manifold with the family $\mathcal{W} = \{W_\alpha, \eta_\alpha\}_{\alpha\in I}$ satisfying conditions* 1 *and* 2. *Then the family $\mathcal{W}^*$ consisting of all pairs which are compatible with $\mathcal{W}$ satisfies conditions* 1, 2, *and* 3, *and therefore makes M into a differentiable manifold.*

Proof. It is clear that, if we prove condition 2 for $\mathcal{W}^*$, we have proved the theorem. Let $\{W, \eta\}$ and $\{W', \eta'\}$ be two pairs which are compatible with $\mathcal{W}$. Let $U = \eta(W)$ and $U' = \eta'(W')$, and assume that $U \cap U'$ is not empty.

Let x be a point of $\eta'^{-1}(U \cap U')$. We must prove that $\eta^{-1}\circ\eta'$ is differentiable at x. By condition 1, there is an index α such that $\eta'(x)$ is contained in $U_\alpha = \eta_\alpha(W_\alpha)$. Hence the set $U_\alpha \cap U \cap U'$ is not empty, and on this set $\eta^{-1}\circ\eta' = (\eta^{-1}\circ\eta_\alpha)\circ(\eta_\alpha^{-1}\circ\eta')$. By the compatibility of $\{W, \eta\}$ with $\mathcal{W}$, we know that the mappings $\eta^{-1}\circ\eta_\alpha$ and $\eta_\alpha^{-1}\circ\eta'$, suitably restricted, are differentiable. Hence $\eta^{-1}\circ\eta'$ is differentiable at x.

We see from Theorem 2-2 and the remarks preceding it that it is sufficient to describe the coordinate neighborhoods of a coordinate manifold in order to define a differentiable manifold. We shall usually be satisfied with such a description when presenting examples of manifolds.

Let M be a differentiable manifold, and let U_α be a coordinate neighborhood of M. Let $\varphi_\alpha\colon U_\alpha \to W_\alpha$ be the associated coordinate mapping. If m is a point in U_α, then $\varphi_\alpha(m) = (x_1, \ldots, x_n)$. In other words, with respect to a particular coordinate neighborhood U_α the coordinate mapping assigns to each point of U_α a set of coordinates $x_1, \ldots, x_n$. When it is clear that the point m is being considered as a point of U_α, we shall speak of it as the point $(x_1, \ldots, x_n)$ instead of $\eta_\alpha(x_1, \ldots, x_n)$. Of course, m may also be a point in another coordinate neighborhood and may have quite different coordinates with respect to that other coordinate neighborhood.

Let M be a differentiable manifold. Let f be a real-valued function defined on an open subset V of M. In order to describe the differentiability of f, we shall refer to the notion of differentiability in E^n by introducing coordinate neighborhoods. Let m be a point in V. Let U_α be a coordinate neighborhood of m, and let $\eta_\alpha\colon W_\alpha \to U_\alpha$ be the corresponding mapping. Then $f\circ\eta_\alpha$ is a function on the set $\eta_\alpha^{-1}(V \cap U_\alpha)$. If $f\circ\eta_\alpha$ is differentiable at $\eta_\alpha^{-1}(m)$, we shall say that f is *differentiable* at m. We must show, of course, that this notion of differentiability does not depend upon the choice of the coordinate neighborhood. Let U_β be another coordinate neighborhood which contains m, and let $\eta_\beta\colon W_\beta \to U_\beta$ be the corresponding mapping. Then

$$f\circ\eta_\beta = (f\circ\eta_\alpha)\circ\psi_\beta{}^\alpha$$

where $\psi_\beta{}^\alpha$ is the coordinate transformation on $U_\alpha \cap U_\beta$. Consequently $f\circ\eta_\beta$ is also differentiable at $\eta_\beta^{-1}(m)$, which was to be proved. If f is differentiable at all points of V, we shall say that f is differentiable on V.

The differentiability of mappings may now easily be defined in terms of the differentiability of functions. Let M and M' be two manifolds. Let $\sigma\colon V \to M'$ be a mapping defined on an open subset V of M. Let m be a point in V. If for every function g which is differentiable on a neighborhood of $\sigma(m)$ the function $g\circ\sigma$ is differentiable at m, we shall say that σ is *differentiable* at m. If σ is differentiable at all points of V, we shall say that σ is differentiable on V.

If δ is a homeomorphism of the manifold M onto itself such that both δ and δ^{-1} are differentiable on M, then δ is called a *diffeomorphism* of M onto itself. It is clear that the set of all such diffeomorphisms forms a group.

EXERCISES

1. Prove that the differentiable varieties in E^n are differentiable manifolds.

2. Prove that the mapping

$$\delta: \quad \begin{aligned} y_1 &= x_1 \sin x_3 + x_2 \cos x_3 \\ y_2 &= x_1 \cos x_3 - x_2 \sin x_3 \\ y_3 &= x_3 \end{aligned}$$

is a diffeomorphism of S^2 (see Section 2-1).

2-4 CONSTRUCTION OF DIFFERENTIABLE MANIFOLDS BY IDENTIFICATION

A topological space M can be given the structure of a differentiable manifold if there is a homeomorphism of M onto a differentiable variety in E^n. For example, any space that is topologically equivalent to the variety in E^3 defined by the equation $x_1^2 + x_2^2 + x_3^2 = 1$ can be made into a differentiable manifold called the 2-sphere. We shall now present some examples of topological spaces which can be given the structure of differentiable manifolds because they are the image of a sufficiently well-behaved mapping defined on E^n. In particular, we shall produce the 2-sphere by this method.

Identify E^3 with V^3 so that the point (0, 0, 0) corresponds to the zero vector. If u is a nonzero vector from V^3, the set H_u of all points (u_1t, u_2t, u_3t) with $t > 0$ will be called the *ray* or half line in E^3 defined by u. If v is another nonzero vector, then $H_v = H_u$ if and only if v and u have the same direction, that is, if and only if there is a positive number c such that $v = cu$.

Let S be the set of all rays of E^3. Let E^* be the set of all points of E^3 that are different from (0, 0, 0). Each point x in E^* belongs to precisely one ray H in E^3. The correspondence $x \to H$ of E^* onto S will be called the *identification mapping* and will be denoted by π. Let U be a subset of S, and denote by $\pi^{-1}(U)$ the set of all points of E^* that are mapped into U by π. We say that U is an open set in S if $\pi^{-1}(U)$ is an open set in E^3. By this definition S becomes a topological space.

Let S^2 be the variety in E^3 defined by the equation

$$x_1^2 + x_2^2 + x_3^2 = 1$$

If H is a point in S, then, as a ray in E^3, H intersects S^2 in precisely one point which we denote by $\sigma(H)$. σ is a one-to-one mapping of S onto S^2 which the reader may easily show is a homeomorphism.

Of course, the homeomorphism σ enables us to introduce the differentiable structure of S^2 into S. But we shall see that the identification mapping can also be used to introduce a differentiable structure into S.

Let f be a real-valued function on an open set U in S. Then $\mathfrak{U} = \pi^{-1}(U)$ is an open set in E^3, and $f^* = f \circ \pi$ is a function on $\mathfrak{U}$. We shall say that f is differentiable on U if f^* is differentiable on $\mathfrak{U}$. With this notion of differentiability for functions, we may define differentiability for mappings in the same manner as for differentiable manifolds.

We shall show that the homeomorphism $\sigma: S \to S^2$, defined previously, and its inverse are differentiable mappings. Let f be a function defined on an open subset U of S^2. Let $g = f \circ \sigma$ and $V = \sigma^{-1}(U)$. Let $\mathfrak{U} = \pi^{-1}(V)$ and $g^* = g \circ \pi$. Then $\mathfrak{U} \cap S^2 = U$ and

$$g^*(x_1, x_2, x_3) = f\left(\frac{x_1}{d}, \frac{x_2}{d}, \frac{x_3}{d}\right)$$

where $d = \sqrt{x_1^2 + x_2^2 + x_3^2}$, for $(x_1, x_2, x_3) \in \mathfrak{U}$. If f is differentiable on U, then clearly g^* is differentiable on $\mathfrak{U}$. By the definition of differentiability which we have adopted for S, this means that g is differentiable on V; hence we have proved that σ is a differentiable mapping. On the other hand, let g be a function defined on an open subset V of S. Let $f = g \circ \sigma^{-1}$ and $U = \sigma(V)$. Let $\mathfrak{U} = \pi^{-1}(V)$ and $g^* = g \circ \pi$. Then $\mathfrak{U} \cap S^2 = U$, and the restriction of g^* to U is the same function as f. If g is differentiable on V, this means that g^* is differentiable on $\mathfrak{U}$. Theorem 2-1 then tells us that f is differentiable on U; hence we have proved that σ^{-1} is differentiable.

It should now be clear that, if we can find a coordinate neighborhood of S with respect to which the differentiable functions are the same as those we have already defined, this coordinate neighborhood will be the image under σ^{-1} of a coordinate neighborhood on S^2. We leave to the reader the verification that S may be given a complete set of coordinate neighborhoods so that it becomes a differentiable manifold with respect to which the differentiable functions are the same as those already defined.

We now have an example of a topological space, namely, the topological 2-sphere, which has defined on it two differentiable structures. These differentiable structures are abstractly the same, in a sense which we shall now make precise. Two differentiable structures M_1 and M_2 on the same topological space X are said to be *isomorphic* if there is a homeomorphism σ of X onto X such that both σ and its inverse σ^{-1} are differentiable mappings.

It is not generally true that two differentiable manifolds that are topologically equivalent must have isomorphic differentiable structures. It has been proved that there are many differentiable structures on the 7-sphere no two of which are isomorphic.

We shall now consider how to give the 2-dimensional torus a differentiable structure. On the one hand, the 2-torus is a surface of revolution in E^3 generated by revolving a circle about a straight line which does not meet it. Consequently it is a differentiable variety in E^3 and can be given the corresponding differentiable structure. We shall now describe another method for producing a differentiable structure on the 2-torus.

Identify V^2 with E^2 so that we may regard E^2 as a group. The group operations induce diffeomorphisms of V^2. Choose a basis $\{v^1, v^2\}$ of V^2. Let D be the subgroup of V^2 consisting of all vectors $a_1v^1 + a_2v^2$, where a_1 and a_2 are integers. Let T denote the quotient group V^2/D. If v is any vector in V^2, the coset $v + D$ will be called the *lattice* in V^2 determined by v. Let $\pi\colon V^2 \to T$ be the natural homomorphism, for which $\pi(v) = v + D$. A subset U of T will be said to be open in T if the subset $\pi^{-1}(U)$ of V^2 is open in V^2. This choice of open sets establishes a topology on T.

Consider the mapping $\sigma^*\colon V^2 \to E^3$ defined by

$$\sigma^*(c_1v^1 + c_2v^2) = ([2 + \cos 2\pi c_1] \cos 2\pi c_2, [2 + \cos 2\pi c_1] \sin 2\pi c_2, \sin 2\pi c_1)$$

The reader may easily verify that $\sigma^*(V^2)$ is the differentiable variety in E^3 which we have called the 2-torus and which we shall denote by T^2. He may also prove that $\sigma^*(u) = \sigma^*(v)$ if and only if u and v are in the same lattice. Consequently σ^* induces a one-to-one mapping σ of T onto T^2 defined by $\sigma(v + D) = \sigma^*(v)$. The mapping σ is a homeomorphism of T onto T^2.

Let f be a function on the open set U in T. Let $\mathfrak{U} = \pi^{-1}(U)$. Then $f \circ \pi$ is a function on $\mathfrak{U}$. We shall say that f is differentiable on U if $f \circ \pi$ is differentiable on $\mathfrak{U}$. In terms of these differentiable functions, the notion of differentiability for mappings into T may be defined in the usual fashion. Just as for the 2-sphere, we find that $\sigma\colon T \to T^2$ is an isomorphism between the two differentiable structures that we have introduced on the 2-torus.

EXERCISES

1. Prove that S is a topological space and that $\sigma: S \to S^2$ is a homeomorphism.

2. Prove that S may be given the structure of a differentiable manifold with respect to which the differentiable functions are the same as those which were defined by means of the identification mapping.

3. Prove that T is a topological space and that $\sigma: T \to T^2$ is a homeomorphism.

4. Prove that T may be given the structure of a differentiable manifold with respect to which the differentiable functions are the same as those which were defined by means of the homomorphism π.

2-5 GERMS OF DIFFERENTIABLE FUNCTIONS

Let M be a differentiable manifold. Let m be a point of M. As in the case of E^n, we may introduce the germs at m of functions which are differentiable near m. Let f_1 and f_2 be two functions, each of which is differentiable on some neighborhood of m. If f_1 and f_2 have the same values on a neighborhood of m, they will be said to have the same *germ* at m. This is an equivalence relation among the functions that are differentiable near m. The equivalence class of all functions that have the same germ as f_1 will be called the germ of f_1.

The set $G^*(m)$ of all germs at m is easily given the structure of a vector space by using the operations of addition and multiplication by numbers on the functions (see Section 1-3). We shall see that $G^*(m)$ is of infinite dimension. Let U be a coordinate neighborhood of m, and let $(x_1, \ldots, x_n)$ be the coordinates of the points of U. We define the sequence of functions $f_0, f_1, f_2, \ldots$ on U by

$$f_j(x_1, \ldots, x_n) = x_1{}^j$$

Suppose that the germs of the functions $f_0, f_1, \ldots, f_r$ were linearly dependent. Then there would be numbers $c_0, \ldots, c_r$, not all zero, such that $c_0f_0 + \cdots + c_rf_r$ would be the zero function on some neighborhood of m. But this would mean that the polynomial $c_0 + c_1x_1 + \cdots + c_rx_1{}^r$ is zero for all values of x_1 in some open interval, which is not true. Thus we have produced an infinite sequence of germs which are linearly independent.

The product h of two functions f and g which are differentiable on a neighborhood of m may be defined by $h(p) = f(p)g(p)$ for points p which are in a neighborhood of m on which both f and g are defined. The function h will then be differentiable on a neighborhood of m. If f_1 and f_2 have the same germ and if g_1 and g_2 have the same germ, then f_1g_1 and f_2g_2 have the same germ since

$$f_1g_1 - f_2g_2 = (f_1 - f_2)g_1 + f_2(g_1 - g_2)$$

Thus we may define the product of the germ of f and the germ of g to be the germ of h, and the result does not depend upon which functions we have used to represent the germs. If we introduce this operation into the vector space $G^*(m)$, then $G^*(m)$ becomes a commutative algebra.

Theorem 2-3. *Let m and m' be two points of a differentiable manifold M. Let $G^*(m)$ and $G^*(m')$ be the corresponding algebras of germs of differentiable functions. Then there is an isomorphism of $G^*(m)$ onto $G^*(m')$.*

Proof. Let U and U' be coordinate neighborhoods of m and m', respectively. Let φ and φ' be the corresponding coordinate mappings. Let τ be the translation of E^n which carries $\varphi'(m')$ onto $\varphi(m)$.

If κ is an element of $G^*(m)$, we may assume that the function f which represents κ is differentiable on a neighborhood V of m which is contained in U. Then $g = f\circ\varphi^{-1}\circ\tau\circ\varphi'$ is a function which is defined on a neighborhood of m'. Since $g\circ\varphi'^{-1} = (f\circ\varphi^{-1})\circ\tau$ is the composition of the differentiable mapping τ and the differentiable function $f\circ\varphi^{-1}$, it follows that $g\circ\varphi'^{-1}$ is differentiable on a neighborhood of $\varphi'(m')$. Thus g is differentiable on a neighborhood of m' and has a germ in $G^*(m')$. We shall call this germ $\sigma(\kappa)$. We must show that this definition depends only upon κ and not upon the function that we have chosen to represent κ.

Let f_1 be another function which represents κ, so that $f - f_1$ is zero on a neighborhood of m. Let $g_1 = f_1\circ\varphi^{-1}\circ\tau\circ\varphi'$. Then

$$g - g_1 = (f - f_1)\circ\varphi^{-1}\circ\tau\circ\varphi'$$

is zero on a neighborhood of m'. Consequently g_1 belongs to the same germ as g, and we have proved that $\sigma\colon G^*(m) \to G^*(m')$ is a well-defined mapping.

Let κ_1 and κ_2 be two elements of $G^*(m)$ which are represented by the functions f_1 and f_2, respectively. If a_1 and a_2 are numbers, then $a_1f_1 + a_2f_2$ represents the germ $a_1\kappa_1 + a_2\kappa_2$ on a suitable neighborhood of m. Hence $\sigma(a_1\kappa_1 + a_2\kappa_2)$ is represented by

$$(a_1f_1 + a_2f_2)\circ\varphi^{-1}\circ\tau\circ\varphi' = a_1(f_1\circ\varphi^{-1}\circ\tau\circ\varphi') + a_2(f_2\circ\varphi^{-1}\circ\tau\circ\varphi')$$

Thus

$$\sigma(a_1\kappa_1 + a_2\kappa_2) = a_1\sigma(\kappa_1) + a_2\sigma(\kappa_2)$$

and we see that $\sigma\colon G^*(m) \to G^*(m')$ is a linear transformation.

In the same manner, we may define a linear mapping σ': $G^*(m') \to G^*(m)$. If g represents the element κ' of $G^*(m')$, then $g \circ \varphi'^{-1} \circ \tau^{-1} \circ \varphi$ represents $\sigma'(\kappa')$ in $G^*(m)$. It is easily seen that $\sigma' \circ \sigma$ and $\sigma \circ \sigma'$ are identity mappings on $G^*(m)$ and $G^*(m')$, respectively. Consequently σ is a one-to-one mapping of $G^*(m)$ onto $G^*(m')$. The reader may easily verify that σ also preserves the operation of multiplication.

Theorem 2-4. *Let M be a differentiable manifold and let $G^*(m)$ be the algebra of germs of differentiable functions at the point m. Then every germ in $G^*(m)$ is represented by a function which is differentiable on all of M.*

This theorem is a consequence of the following lemma.

Lemma 2-1. *Let $0 < r_1 < r_2$. Let $|x|$ denote the distance of a point x of E^n from $(0, \ldots, 0)$. Then there is a differentiable function h on E^n such that*

$$h(x) = \begin{cases} 1 & \text{for } |x| \leqq r_1 \\ 0 & \text{for } |x| \geqq r_2 \end{cases}$$

Proof. Define the functions $q_1(t)$ and $q_2(t)$ by

$$q_1(t) = \begin{cases} 0 & \text{for } 0 \leqq t \leqq r_1 \\ \exp\{-(t - r_1)^{-2}\} & \text{for } \quad t > r_1 \end{cases}$$

$$q_2(t) = \begin{cases} \exp\{-(t - r_2)^{-2}\} & \text{for } 0 \leqq t < r_2 \\ 0 & \text{for } \quad t \geqq r_2 \end{cases}$$

Then q_1 and q_2 are differentiable for $t \geqq 0$; hence $q(t) = q_1(t)q_2(t)$ is differentiable for $t \geqq 0$. Furthermore, the integral $C = \int_0^\infty q(t)\,dt$ is finite. Define

$$h(x) = \frac{1}{C} \int_{|x|}^{\infty} q(t)\,dt$$

Since

$$\frac{\partial h}{\partial x_j} = -\frac{1}{C}\, q(|x|)\, \frac{x_j}{|x|}$$

we see that h is differentiable, and it clearly satisfies the other desired conditions.

Proof of Theorem 2-4. Let f be a function on the manifold M which is differentiable on a neighborhood of m. We may assume that f is differentiable on the coordinate neighborhood U of m with the coordinates $(x_1, \ldots, x_n)$ and that m has the coordinates $(0, \ldots, 0)$. There is a number $r_2 > 0$ such that the points x of E^n which satisfy $|x| \leqq r_2$ give coordinates of points which are in U. Choose r_1 so that $0 < r_1 < r_2$. Let h be the function described in Lemma 2-1. Define

$f^*(x) = f(h(x)x_1, \ldots, h(x)x_n)$ for the points of U, and define

$$f^*(m') = f(0, \ldots, 0)$$

for those points m' of M which lie outside U. Then f^* is differentiable on all of M. Since $f^*(x) = f(x)$ for $|x| < r_1$, we see that f^* has the same germ as f at m.

Theorem 2-4 shows that for the investigation of the local properties of a differentiable manifold it involves no loss of generality to assume that the functions used are differentiable on all of M.

EXERCISE

Define the germ of a differentiable mapping σ: $M \to M'$ at a point m of M. Prove that every germ of a differentiable mapping may be represented by a mapping which is differentiable on all of M.

2-6 THE TANGENT AND COTANGENT SPACES AT A POINT

Let M be a differentiable manifold, and let m be a point of M. Let $\{U_\alpha, \varphi_\alpha\}$ be the indexed family of all coordinate neighborhoods of M *which contain the point* m and their corresponding coordinate mappings. For each pair of indices α and β we may introduce the coordinate transformation $\psi_\alpha^\beta = \varphi_\beta \circ \varphi_\alpha^{-1}$ defined on $\varphi_\alpha(U_\alpha \cap U_\beta)$. In this way we obtain the doubly indexed family $\{\psi_\alpha^\beta\}$ of mappings. If γ is a third index, then

$$\psi_\beta^\gamma \circ \psi_\alpha^\beta = \varphi_\gamma \circ \varphi_\beta^{-1} \circ \varphi_\beta \circ \varphi_\alpha^{-1} = \varphi_\gamma \circ \varphi_\alpha^{-1} = \psi_\alpha^\gamma$$

on $\varphi_\alpha(U_\alpha \cap U_\beta \cap U_\gamma)$. The relations $\psi_\beta^\gamma \circ \psi_\alpha^\beta = \psi_\alpha^\gamma$ will be called the *coherence relations* of the family of coordinate transformations at m.

The reader should bear in mind that the domain and range of the mapping ψ_α^β are subsets of E^n, so that ψ_α^β has the explicit form

$$\psi_\alpha^\beta: \quad \begin{aligned} y_1 &= g_1(x_1, \ldots, x_n) \\ &\cdots\cdots\cdots \\ y_n &= g_n(x_1, \ldots, x_n) \end{aligned}$$

where $(x_1, \ldots, x_n)$ ranges over the points of the set $\varphi_\alpha(U_\alpha \cap U_\beta)$. The numbers $x_1, \ldots, x_n$ may also be looked upon as coordinates of the point $\varphi_\alpha^{-1}(x_1, \ldots, x_n)$ in U_α and the numbers $y_1, \ldots, y_n$ as coordinates of the *same point* $\varphi_\beta^{-1}(y_1, \ldots, y_n)$ in U_β if

$$(y_1, \ldots, y_n) = \psi_\alpha^\beta(x_1, \ldots, x_n)$$

It is then a question of two systems of coordinates for the points of $U_\alpha \cap U_\beta$, and $\psi_\alpha{}^\beta$ describes the relationship of one system of coordinates to the other.

Let $x^\circ = \varphi_\alpha(m)$ and $y^\circ = \varphi_\beta(m)$, so that $y^\circ = \psi_\alpha{}^\beta(x^\circ)$. The differentiable mapping $\psi_\alpha{}^\beta$: $\varphi_\alpha(U_\alpha \cap U_\beta) \to \varphi_\beta(U_\alpha \cap U_\beta)$ induces the linear mapping $\psi_\alpha{}^\beta{}_*: T(x^\circ) \to T(y^\circ)$ of the tangent space to E^n at x° into the tangent space at y°. The coherence relations $\psi_\beta{}^\gamma \circ \psi_\alpha{}^\beta = \psi_\alpha{}^\gamma$ imply the coherence relations $\psi_\beta{}^\gamma{}_* \circ \psi_\alpha{}^\beta{}_* = \psi_\alpha{}^\gamma{}_*$ for the induced linear mappings. Furthermore, $\psi_\alpha{}^\alpha{}_*$ is the identity mapping, so that $\psi_\alpha{}^\beta{}_*$ is nonsingular and $\psi_\beta{}^\alpha{}_*$ is its inverse. The family of coordinate neighborhoods of m therefore produces a family of n-dimensional vector spaces which are linked together by the coherence relations

$$\psi_\beta{}^\gamma{}_* \circ \psi_\alpha{}^\beta{}_* = \psi_\alpha{}^\gamma{}_*$$

In a similar fashion, $\psi_\alpha{}^\beta$ also induces a nonsingular mapping $\psi_\alpha{}^{\beta *}$ of the cotangent space $T^*(y^\circ)$ to E^n onto the cotangent space $T^*(x^\circ)$. The resulting family of n-dimensional vector spaces is interrelated by the coherence relations $\psi_\alpha{}^{\beta *} \circ \psi_\beta{}^{\gamma *} = \psi_\alpha{}^{\gamma *}$.

Let $\mathfrak{J}(m)$ be the set of all pairs (x, α), where $x = \varphi(m)$ for some coordinate mapping φ and α is a tangent vector to E^n at x. The symbol α which denotes a tangent vector to E^n should not be confused with the index α for coordinate neighborhoods. Let (x, α) and (x', α') be two pairs from $\mathfrak{J}(m)$. We shall say that (x, α) is equivalent to (x', α'), or, less precisely, that α and α' represent the same tangent vector at m, when $x' = \psi(x)$ and $\alpha' = \psi_*(\alpha)$ for some coordinate transformation ψ at m. It is clear from the coherence relations that this is an equivalence relation in $\mathfrak{J}(m)$. The equivalence classes of pairs will be called *tangent vectors* to M at m. We shall denote the equivalence class that contains the pair (x, α) by $\{\alpha\}$.

In a similar fashion, we may define differentials on M at m. Let $\mathfrak{J}^*(m)$ be the set of all pairs (x, λ), where $x = \varphi(m)$ for some coordinate mapping φ and λ is a differential at x. Two pairs (x, λ) and (x', λ') from $\mathfrak{J}^*(m)$ will be called equivalent, or we shall say that λ and λ' represent the same differential at m, when $x' = \psi(x)$ and $\lambda = \psi^*(\lambda')$ for some coordinate transformation ψ at m. We shall denote the equivalence class that contains the pair (x, λ) by $\{\lambda\}$.

We may introduce vector operations among the tangent vectors to M at m. Suppose that (x, α) and (x', α') are equivalent elements of $\mathfrak{J}(m)$, so that $x' = \psi(x)$ and $\alpha' = \psi_*(\alpha)$. Let β be a second tangent vector to E^n at x and $\beta' = \psi_*(\beta)$. Since ψ_* is a linear mapping, it follows that

$$a\alpha' + b\beta' = \psi_*(a\alpha + b\beta)$$

where a and b are numbers, and consequently $(x, a\alpha + b\beta)$ is equivalent to $(x', a\alpha' + b\beta')$. We may therefore legitimately define

$$a\{\alpha\} + b\{\beta\} = \{a\alpha + b\beta\}$$

Under this definition, the set $T(m)$ of all tangent vectors to M at m becomes a vector space with the same algebraic structure as the tangent space to E^n at x. $T(m)$ will be called the tangent space to M at m. The cotangent space $T^*(m)$ to M at m may be defined similarly, and it will have the algebraic structure of the cotangent space to E^n at x.

Let $\{\alpha\}$ be a tangent vector to M at m and $\{\lambda\}$ a differential at M. We may assume that the representatives α and λ are associated with the same coordinate neighborhood U. If ψ is a coordinate transformation at m from U to another coordinate neighborhood, then

$$\langle \psi_*(\alpha), \psi^{*-1}(\lambda)\rangle = \langle \alpha, \lambda\rangle$$

Consequently we may define

$$\langle \{\alpha\}, \{\lambda\}\rangle = \langle \alpha, \lambda\rangle$$

and the value will not depend upon the choice of the coordinate neighborhood that we have used to define it. It follows immediately that the rules for the "product" $\langle \{\alpha\}, \{\lambda\}\rangle$ are the same as those for $\langle \alpha, \lambda\rangle$ and hence that $T(m)$ and $T^*(m)$ are dual spaces to one another.

At this point the reader may reasonably protest that a tangent vector to M ought to be associated with certain curves in M and that a differential on M ought to be the differential of a function on M. We shall show how this may be brought about.

Let f be a function which is differentiable on a neighborhood of the point m of the manifold M. Let U be a coordinate neighborhood of the point m and $\varphi: U \to W$ the corresponding coordinate mapping. Then $f \circ \varphi^{-1}$ is a function which is differentiable on a neighborhood of the point $\varphi(m)$ in E^n. If U' is another such coordinate neighborhood and φ' is its coordinate mapping, then

$$f \circ \varphi^{-1} = f \circ \varphi'^{-1} \circ \psi$$

where ψ is the coordinate transformation on $U \cap U'$. Consequently the differentials of these functions on E^n satisfy

$$\psi^*(d(f \circ \varphi'^{-1})) = d(f \circ \varphi^{-1})$$

and we see that $d(f \circ \varphi'^{-1})$ and $d(f \circ \varphi^{-1})$ represent the same differential at m on M. This differential $\{d(f \circ \varphi^{-1})\}$ will be denoted by df since it does not depend upon the choice of the coordinate neighborhood that we used to define it. Since

$$d[(a_1f_1 + a_2f_2)\circ\varphi^{-1}] = a_1d(f_1\circ\varphi^{-1}) + a_2d(f_2\circ\varphi^{-1})$$

it follows that

$$d(a_1f_1 + a_2f_2) = a_1df_1 + a_2df_2$$

If the pair $(\varphi(m), \lambda)$ represents the differential $\{\lambda\}$ on M at m, then $\lambda = dg$ for some function g on E^n which is differentiable on a neighborhood of $\varphi(m)$. Let $f = g\circ\varphi$. Then f is a function on M which is differentiable on a neighborhood of m. Since $g = f\circ\varphi^{-1}$, it follows that $\{\lambda\} = df$. We have thus demonstrated that every differential at m is the differential of some function on M.

Let J be the interval $-1 < t < 1$ and $\bar{J}$ its closure in E^1. A continuous mapping $\gamma: \bar{J} \to M$ for which the restriction to J is differentiable is called a differentiable parametrized curve in M. If $\gamma(0) = m$, then γ will be said to be a curve through m. If γ passes through m and if f is a function which is differentiable on a neighborhood of m, we may define

$$\langle\gamma, f\rangle = \frac{d}{dt} f\circ\gamma \bigg|_{t=0}$$

We would like to define $\langle\gamma, df\rangle = \langle\gamma, f\rangle$. In order to justify this definition, we must show that, if g is another function which is differentiable on a neighborhood of m and if $dg = df$, then $\langle\gamma, g\rangle = \langle\gamma, f\rangle$. Let U be a coordinate neighborhood of m and φ the corresponding coordinate mapping. Then $f\circ\gamma = (f\circ\varphi^{-1})\circ(\varphi\circ\gamma)$ and $g\circ\gamma = (g\circ\varphi^{-1})\circ(\varphi\circ\gamma)$ for small values of t. To say that $dg = df$ is to say that $d(g\circ\varphi^{-1}) = d(f\circ\varphi^{-1})$. Hence

$$\frac{d}{dt}[(g\circ\varphi^{-1})\circ(\varphi\circ\gamma)]\bigg|_{t=0} = \frac{d}{dt}[(f\circ\varphi^{-1})\circ(\varphi\circ\gamma)]\bigg|_{t=0}$$

and thus

$$\frac{d}{dt} g\circ\gamma \bigg|_{t=0} = \frac{d}{dt} f\circ\gamma \bigg|_{t=0}$$

or

$$\langle\gamma, g\rangle = \langle\gamma, f\rangle$$

It now follows that, for a given differentiable parametrized curve through m, the mapping $df \to \langle\gamma, df\rangle$ is a linear functional on $T^*(m)$ and corresponds to a unique element $\{\alpha\}$ of the dual space $T(m)$. The tangent vector $\{\alpha\}$ will be called the tangent vector to γ at m. Two differentiable parametrized curves γ_1 and γ_2 through m will have the same tangent vector if and only if $\langle\gamma_1, \{\lambda\}\rangle = \langle\gamma_2, \{\lambda\}\rangle$ for all differentials $\{\lambda\}$.

EXERCISE

Let S be a differentiable variety in E^n. Any differentiable parametrized curve in S may also be regarded as a differentiable parametrized curve in E^n. Prove that two differentiable parametrized curves in S have the same tangent vector at a point if and only if they have the same tangent vector as curves in E^n.

2-7 DIFFERENTIABLE MAPPINGS AND THEIR INDUCED LINEAR TRANSFORMATIONS

Let M and $\bar{M}$ be two differentiable manifolds. Let V be an open subset of M, and let $\sigma: V \to \bar{M}$ be a differentiable mapping. Let $m \in V$. Let T be the tangent space to M at m and $\bar{T}$ the tangent space to $\bar{M}$ at $\sigma(m)$. We wish to derive a linear transformation $\sigma_*: T \to \bar{T}$ from the mapping $\sigma: V \to \bar{M}$. Naturally we must examine the linear transformations of the tangent spaces to E^n that are induced by σ when we introduce coordinate neighborhoods of m and $\sigma(m)$.

Let U be a coordinate neighborhood of m in M and $\bar{U}$ a coordinate neighborhood of $\sigma(m)$ in $\bar{M}$. Let φ and $\bar{\varphi}$ be the corresponding coordinate mappings. Then $\sigma: V \to \bar{M}$ induces a differentiable mapping $\sigma_1: \varphi(U \cap \sigma^{-1}(\bar{U}) \cap V) \to \bar{\varphi}(\bar{U})$ given by $\sigma_1 = \bar{\varphi}\circ\sigma\circ\varphi^{-1}$ which, in turn, induces a linear transformation $\Sigma(U, \bar{U}) = \sigma_{1*}$ of the tangent space T_U to E^n at $\varphi(m)$ into the tangent space $\bar{T}_{\bar{U}}$ to E^n at $\bar{\varphi}(\sigma(m))$. Briefly, each pair $U, \bar{U}$ of suitable coordinate neighborhoods gives rise in a natural way to a linear transformation $\Sigma(U, \bar{U}): T_U \to \bar{T}_{\bar{U}}$. We must examine the relationships among the linear transformations $\Sigma(U, \bar{U})$.

Let $U', \bar{U}'$ be another pair of suitable coordinate neighborhoods and $\varphi', \bar{\varphi}'$ the corresponding coordinate mappings. Let $\psi = \varphi'\circ\varphi^{-1}$ and $\bar{\psi} = \bar{\varphi}'\circ\bar{\varphi}^{-1}$ be the corresponding coordinate transformations. From $\sigma_1' = \bar{\varphi}'\circ\sigma\circ\varphi'^{-1} = \bar{\psi}\circ\sigma_1\circ\psi^{-1}$ we obtain

$$\Sigma(U', \bar{U}') = \bar{\psi}_*\circ\Sigma(U, \bar{U})\circ\psi_*^{-1} \tag{2-7-1}$$

Our problem is to derive a well-defined linear transformation $\sigma_*: T \to \bar{T}$ from the family of mappings $\Sigma(U, \bar{U})$.

Let θ_* be the natural isomorphism of T_U onto T given by $\theta_*(\alpha) = \{\alpha\}$. Let $\bar{\theta}_*: \bar{T}_{\bar{U}} \to \bar{T}$ be the natural isomorphism. Then we obtain a linear transformation $\sigma_*: T \to \bar{T}$ by defining $\sigma_* = \bar{\theta}_*\circ\Sigma(U, \bar{U})\circ\theta_*^{-1}$. If we change from the pair $U, \bar{U}$ to the pair $U', \bar{U}'$, the natural isomorphisms

$\theta'_*: T_{U'} \to T$ and $\bar{\theta}'_*: \bar{T}_{\bar{U}'} \to T$ satisfy $\theta'_* = \theta_* \circ \psi_*^{-1}$ and $\bar{\theta}'_* = \bar{\theta}_* \circ \bar{\psi}_*^{-1}$. Hence, using the relation (2-7-1),

$$\bar{\theta}'_* \circ \Sigma(U', \bar{U}') \circ \theta'^{-1}_* = \bar{\theta}_* \circ \Sigma(U, \bar{U}) \circ \theta_*^{-1}$$

and we see that the definition of σ_* is independent of the choice of the pair U, $\bar{U}$ of coordinate neighborhoods. We have therefore proved that $\sigma: V \to \bar{M}$ induces in a natural way a linear transformation on the tangent space to M at each point of V.

In a similar fashion, $\sigma: V \to \bar{M}$ also induces a linear transformation $\sigma^*: \bar{T}^* \to T^*$, where $\bar{T}^*$ is the cotangent space to $\bar{M}$ at $\sigma(m)$ and T^* is the cotangent space to M at m. We shall leave the details of the construction of σ^* to the reader.

In the proof of the next theorem we shall see how the behavior of the differentiable mapping $\sigma: V \to \bar{M}$ on a neighborhood of the point m is reflected in the behavior of the induced linear transformation $\sigma_*: T \to \bar{T}$. Roughly speaking, $\sigma: V \to \bar{M}$ is one-to-one on a neighborhood of m if and only if $\sigma_*: T \to \bar{T}$ is one-to-one, and σ maps suitably small neighborhoods of m onto open sets if and only if σ_* is onto.

If the induced linear transformation $\sigma_*: T \to \bar{T}$ is one-to-one at a point m, the differentiable mapping $\sigma: V \to \bar{M}$ is said to be *regular* at m. If σ is regular at each point of V, then σ is said to be regular on V.

Theorem 2-5. *Let V be an open subset of a differentiable manifold M. Let $\bar{M}$ be another differentiable manifold, and let $\sigma: V \to \bar{M}$ be a differentiable mapping. Let m be a point of V. Then σ is regular at m if and only if there are a neighborhood W of $\sigma(m)$ and a differentiable mapping $\tau: W \to V$ such that $\tau(W)$ is an open set and $\tau \circ \sigma$ is the identity mapping on $\tau(W)$.*

Proof. Since we may assume that the neighborhoods on which our mappings are defined are coordinate neighborhoods, it suffices to prove this theorem when M and $\bar{M}$ are euclidean spaces. Thus we may assume that the mapping σ is given by

$$\sigma: \quad \begin{array}{l} y_1 = g_1(x_1, \ldots, x_n) \\ \cdots\cdots\cdots\cdots \\ y_r = g_r(x_1, \ldots, x_n) \end{array}$$

Then
$$\sigma_*\left(\frac{\partial}{\partial x_i}\right) = \sum_{j=1}^{r} \frac{\partial g_j}{\partial x_i}\left(\frac{\partial}{\partial y_j}\right)$$

for $i = 1, \ldots, n$.

Let σ be regular at m. We shall prove that the columns of the $r \times n$ matrix

$$\mathcal{J} = \begin{pmatrix} \dfrac{\partial g_1}{\partial x_1} & \cdots & \dfrac{\partial g_1}{\partial x_n} \\ \cdots & \cdots & \cdots \\ \dfrac{\partial g_r}{\partial x_1} & \cdots & \dfrac{\partial g_r}{\partial x_n} \end{pmatrix}$$

are linearly independent at the point m. Let $c_1, \ldots, c_n$ be numbers such that

$$\sum_{i=1}^{n} c_i \frac{\partial g_j}{\partial x_i}\bigg|_m = 0$$

for $j = 1, \ldots, r$. Then at m

$$\sigma_*\left(\sum_{i=1}^{n} c_i \frac{\partial}{\partial x_i}\right) = \sum_{i=1}^{n} c_i \sigma_* \left(\frac{\partial}{\partial x_i}\right) = \sum_{j=1}^{r}\sum_{i=1}^{n} c_i \frac{\partial g_j}{\partial x_i}\left(\frac{\partial}{\partial y_j}\right) = 0$$

Since σ_* is one-to-one at m, it follows that

$$\sum_{i=1}^{n} c_i \left(\frac{\partial}{\partial x_i}\right)_m = 0$$

and hence that $c_1 = \cdots = c_n = 0$.

Since the matrix $\mathcal{J}$ has rank n, there are subscripts $j_1, \ldots, j_n$ such that the $n \times n$ matrix

$$\mathcal{J}_1 = \begin{pmatrix} \dfrac{\partial g_{j_1}}{\partial x_1} & \cdots & \dfrac{\partial g_{j_1}}{\partial x_n} \\ \cdots & \cdots & \cdots \\ \dfrac{\partial g_{j_n}}{\partial x_1} & \cdots & \dfrac{\partial g_{j_n}}{\partial x_n} \end{pmatrix}$$

is nonsingular at m. We may apply the implicit-function theorem to the relations

$$\begin{aligned} g_{j_1}(x_1, \ldots, x_n) - y_{j_1} &= 0 \\ \cdots\cdots\cdots\cdots\cdots & \\ g_{j_n}(x_1, \ldots, x_n) - y_{j_n} &= 0 \end{aligned} \tag{2-7-2}$$

Let $m = (x_1^\circ, \ldots, x_n^\circ)$. Let $\sigma(m) = y^\circ = (y_1^\circ, \ldots, y_r^\circ)$ and $z^\circ = (y_{j_1}^\circ, \ldots, y_{j_n}^\circ)$. Then the relations (2-7-2) are satisfied by the point (m, z°) in E^{2n}. Using the implicit-function theorem, we may assert that there are a neighborhood U of z° in E^n and a differentiable mapping $\rho\colon U \to V$ such that $\rho(z^\circ) = m$ and

$$g_{j_1}(\rho(z)) - z_1 = 0$$
$$\cdots\cdots\cdots\cdots$$
$$g_{j_n}(\rho(z)) - z_n = 0$$

for z in U. Let $\pi\colon E^r \to E^n$ be the projection for which

$$\pi(y_1, \ldots, y_r) = (y_{j_1}, \ldots, y_{j_n})$$

In terms of π, our assertion is that $\pi\circ\sigma\circ\rho$ is the identity mapping on U.

For the induced linear transformations on the tangent spaces we must have $(\pi\circ\sigma)_*\circ\rho_* =$ identity mapping. Hence ρ is regular at the point z°. We may repeat for ρ the argument which we have just presented for σ. Thus there are a neighborhood V_1 of m, which we may assume is contained in V, and a differentiable mapping $\theta\colon V_1 \to U$ such that $\theta(m) = z^\circ$ and $\rho\circ\theta$ is the identity mapping on V_1 (see Figure 2-4).

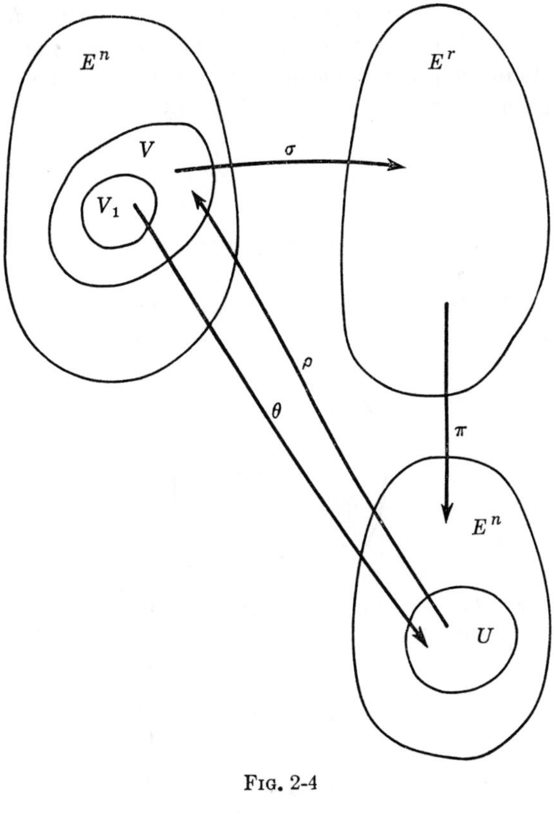

FIG. 2-4

On V_1 we have $\pi\circ\sigma = \pi\circ\sigma\circ\rho\circ\theta = \theta$. Hence $\theta\circ\rho$ is the identity mapping on $\theta(V_1)$. If u is a point of U, $\rho(u)$ is in V_1, and $u' = \theta\circ\rho(u)$, then $\rho(u') = \rho(u)$. Hence $\pi\circ\sigma\circ\rho(u') = \pi\circ\sigma\circ\rho(u)$ or $u' = u$. This proves that u is in $\theta(V_1)$ and hence that $\rho^{-1}(V_1) \subset \theta(V_1)$. But then $\theta(V_1) = \rho^{-1}(V_1)$; hence $\theta(V_1)$ is an open set in E^n.

Let $W = \pi^{-1}[\theta(V_1)]$. Then W is an open set. Define $\tau = \rho\circ\pi$: $W \to V$. Then $\tau(W) = \rho\circ\pi(W) = \rho\circ\theta(V_1) = V_1$ is an open set. On $\tau(W) = V_1$ we find that $\tau\circ\sigma = \rho\circ\pi\circ\sigma = \rho\circ\theta =$ identity mapping.

The converse statement follows because $\tau\circ\sigma =$ identity implies $\tau_*\circ\sigma_* =$ identity for the induced linear transformations; hence σ_* is one-to-one.

Theorem 2-6. *Let V be an open subset of a differentiable manifold M. Let $\bar{M}$ be another differentiable manifold, and let σ: $V \to \bar{M}$ be a differentiable mapping. Let $m \in V$, and let σ_*: $T \to \bar{T}$ be the linear transformation that σ induces on the tangent space to M at m. Then σ_* is onto if and only if there are a neighborhood W of $\sigma(m)$ and a differentiable mapping τ: $W \to V$ such that $\sigma\circ\tau$ is the identity mapping on W.*

Proof. As in the proof of Theorem 2-5, we may assume that M and $\bar{M}$ are euclidean spaces and that σ is given by

$$\sigma: \quad \begin{aligned} y_1 &= g_1(x_1, \ldots, x_n) \\ &\cdots\cdots\cdots\cdots \\ y_r &= g_r(x_1, \ldots, x_n) \end{aligned}$$

Let σ_* be onto. By renumbering the variables $x_1, \ldots, x_n$ we may assume that $\{\sigma_*(\partial/\partial x_1), \ldots, \sigma_*(\partial/\partial x_r)\}$ is a basis for $\bar{T}$. Thus the $r \times r$ matrix

$$\begin{pmatrix} \dfrac{\partial g_1}{\partial x_1} & \cdots & \dfrac{\partial g_1}{\partial x_r} \\ \cdots & \cdots & \cdots \\ \dfrac{\partial g_r}{\partial x_1} & \cdots & \dfrac{\partial g_r}{\partial x_r} \end{pmatrix}$$

is nonsingular. Let $m = (x_1^\circ, \ldots, x_n^\circ)$ and $\sigma(m) = (y_1^\circ, \ldots, y_r^\circ)$.

We may apply the implicit-function theorem to the relations

$$\begin{aligned} g_1(x_1, \ldots, x_n) - y_1 &= 0 \\ &\cdots\cdots\cdots\cdots \\ g_r(x_1, \ldots, x_n) - y_r &= 0 \end{aligned}$$

There are a neighborhood W_1 of $(x^\circ_{r+1}, \ldots, x_n^\circ, y_1^\circ, \ldots, y_r^\circ)$ in E^n and a differentiable mapping τ_1: $W_1 \to U$ such that

$$\tau_1(x^\circ_{r+1}, \ldots, x_n^\circ, y_1^\circ, \ldots, y_r^\circ) = (x_1^\circ, \ldots, x_r^\circ)$$

and

$$g_1(\tau_1(x_{r+1}, \ldots, x_n, y), x_{r+1}, \ldots, x_n) - y_1 = 0$$
$$\cdots\cdots\cdots\cdots\cdots\cdots\cdots\cdots$$
$$g_r(\tau_1(x_{r+1}, \ldots, x_n, y), x_{r+1}, \ldots, x_n) - y_r = 0$$

Let W be the neighborhood of y° in E^r which consists of those points y such that $(x^\circ_{r+1}, \ldots, x_n^\circ, y) \in W_1$. Define the differentiable mapping τ: $W \to V$ by

$$\tau(y) = (\tau_1(x^\circ_{r+1}, \ldots, x_n^\circ, y), x^\circ_{r+1}, \ldots, x_n^\circ)$$

It follows that $\sigma \circ \tau$ = identity on W.

The converse follows by a simple argument like that in the proof of Theorem 2-5.

EXERCISES

Let σ: $V \to \bar{M}$ be a differentiable mapping. Let $m \in V$. Prove the following:

1. If $\bar{f}$ is a differentiable function defined on a neighborhood of $\sigma(m)$, then $\sigma^*(d\bar{f}) = d(\bar{f} \circ \sigma)$.

2. If γ is a differentiable parametrized curve through m and if $\{\alpha\}$ is the tangent vector to γ at m, then $\sigma_*(\{\alpha\})$ is the tangent vector to $\sigma \circ \gamma$ at $\sigma(m)$.

3. σ is regular if and only if σ^* is onto.

4. σ_* is onto if and only if σ^* is one-to-one.

CHAPTER 3

Projective Spaces and Projective Algebraic Varieties

The projective spaces and the nonsingular projective algebraic varieties are examples of abstract differentiable manifolds whose natural development does not display them as differentiable varieties in a euclidean space. Many of the details of their presentation will be left to the reader as exercises, although not explicitly stated as such. It is hoped that the reader will gain more insight into the nature of these examples by being forced to consider for himself the details of their construction.

3-1 THE PROJECTIVE PLANE

The projective plane is, perhaps, the earliest example of an abstract differentiable manifold in mathematical history. We shall give a heuristic account of this concept, roughly following its historical development.

We begin with the affine plane A^2. The incidence relations in A^2 are such that two points are contained in precisely one line, but two lines either intersect in precisely one point or are parallel. The projective plane is obtained by adjoining more points to A^2 in such a way that every pair of lines intersects in precisely one point and every pair of points, including the new ones, is contained in precisely one line.

Let l be a line in A^2. Let p be the family of all lines in A^2 that are parallel to l (including l). p will be called an *ideal point* or *point at infinity*. In the geometry of the projective plane, p is an additional point of the line l. It follows that all lines of the family p have the common point of intersection p. Two lines which are not parallel have distinct points at infinity and so they intersect only at an ordinary point. In order to ensure that two points at infinity are contained in precisely one line, it is agreed that the points at infinity shall constitute a line called the *ideal line* or the *line at infinity.* The set obtained from A^2 by the adjunction of the ideal points of the ideal line with the incidence relations which we have just described will be called the *projective plane* and will be denoted by P^2.

The incidence relations in P^2 may be summarized as follows:

1. Two ordinary lines which are not parallel meet in an ordinary point.

2. Two ordinary lines which are parallel meet in their common ideal point.

3. An ordinary line and the ideal line meet in that ideal point which is on the ordinary line.

4. Two ordinary points are contained in an ordinary line.

5. An ordinary point and an ideal point are contained in an ordinary line, the one line in the family of parallel lines which passes through the given ordinary point.

6. Two ideal points are contained in the ideal line.

The duality in the incidence relations in P^2 has been achieved at a price. Apparently we must deal with two different kinds of lines and two different kinds of points. One of the great achievements of mathematics was the demonstration that these distinctions are only apparent, in the following sense. Let $\tau\colon P^2 \to P^2$ be a one-to-one correspondence of the projective plane onto itself. If l is a line in P^2, the image $\tau(l)$ is a set of points in P^2. If $\tau(l)$ is a line for every line l in P^2, then τ will be called a *projective transformation.* Such a transformation preserves the incidence relations among lines and points, and so it does not alter the geometry of lines and points in P^2. We shall ultimately prove the following theorem: *Given any two lines l and m (any two points p and q) in P^2, there exists a projective transformation which carries l onto m (p onto q).* This remarkable statement shows that it is meaningless to distinguish between ordinary and ideal lines and between ordinary and ideal points in P^2. They play the same role in the geometry.

The view of the projective plane which we have taken so far fails to give it the structure of a differentiable manifold or even a topology. We shall show that there is one and only one topology which may be assigned to P^2 in a "natural" way, that is, so as to satisfy the following two conditions:

1. The topology of P^2 induces on A^2, as a subset of P^2, the usual topology.

2. Every projective transformation of P^2 is a homeomorphism.

There are two general methods by which we may prove this statement: the synthetic approach, which proceeds in the spirit of the preceding discussion, and the analytic approach, which makes use of a numerical model for P^2. The analytic approach is the one which we shall adopt since it gives us the desired results with greater ease.

Identify A^2 with E^2. Let S be a sphere with center c which is tangent to E^2. If p is an ordinary point of P^2 and thus a point in E^2, the line through p and c intersects S in precisely two diametrically opposite points p' and p'' (see Figure 3-1). If we regard S as the sphere $x_1^2 + x_2^2 + x_3^2 = 1$ in E^3, then p' and p'' have coordinates (a_1, a_2, a_3) and $(-a_1, -a_2, -a_3)$, with $a_3 \neq 0$. An ordinary line in P^2,

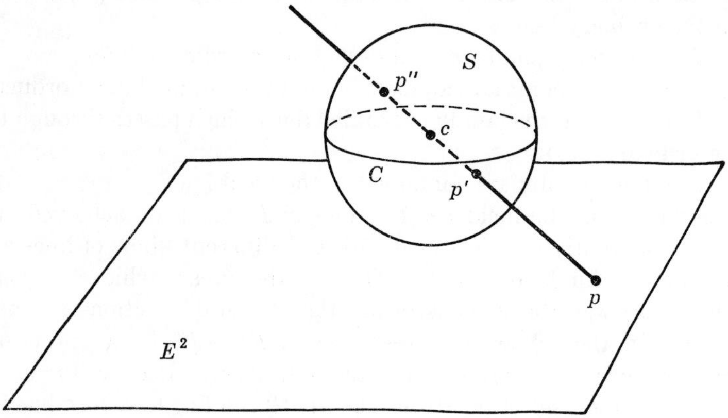

FIG. 3-1

which is a line in E^2, projects onto a great circle of S with the exception of the two points where this great circle intersects the equatorial circle C. Every great circle of S except C is the projection of some line in E^2. If l_1 and l_2 are two parallel lines in E^2, the projections of l_1 and l_2 on S (as great circles on S) intersect in two diametrically opposite points on the great circle C. Indeed, every line parallel to l_1 projects into a great circle which passes through these same two points. Consequently we shall regard these two points as the images of the point at infinity on l_1. Distinct points at infinity correspond to distinct pairs of diametrically opposite points on C. The coordinates of the points on C have the form $(a_1, a_2, 0)$.

By means of the central projection just described, the points of P^2, including the ideal points, are in one-to-one correspondence with the pairs of diametrically opposite points on S. In order to assign coordinates to the points of P^2, it is more convenient to associate with each point p of P^2, including the ideal points, the line through p' and p''. In this way we obtain a one-to-one correspondence between the points of P^2 and the lines in E^3 which pass through (0, 0, 0). We may then assign to the point p any one of the sets of coordinates (a_1, a_2, a_3) of the points on the line through p' and p'' *except* (0, 0, 0). It follows that two sets (a_1, a_2, a_3) and (b_1, b_2, b_3) of coordinates are assigned to the same point of P^2 if and only if there is a nonzero number c such that

$$a_1 = cb_1 \qquad a_2 = cb_2 \qquad a_3 = cb_3$$

The preceding sketch indicates how we should proceed formally to define a numerical model for P^2, and we shall now carry out this procedure. Let E^* be the set of all triples (x_1, x_2, x_3) in R^3 except (0, 0, 0). Two triples (x_1, x_2, x_3) and (y_1, y_2, y_3) are called *equivalent* if there is a number c such that

$$x_1 = cy_1 \qquad x_2 = cy_2 \qquad x_3 = cy_3$$

The relation of equivalence in E^* partitions it into mutually disjoint equivalence classes. The family of these equivalence classes will be denoted by P'. The elements of P' will be called *points*. If (x_1, x_2, x_3) belongs to a point of P', then (x_1, x_2, x_3) will be called a set of *coordinates* of that point.

We must also define lines in P'. Let a_1, a_2, a_3 be real numbers which are not all zero. Let p' be a point in P' and (x_1, x_2, x_3) be coordinates of p'. If these coordinates satisfy the relation

$$a_1x_1 + a_2x_2 + a_3x_3 = 0$$

then every set of coordinates of p' satisfies this relation. In this case we say that the point p' satisfies the relation

$$a_1x_1 + a_2x_2 + a_3x_3 = 0 \tag{3-1-1}$$

The set of all points of P' that satisfy a relation of the form (3-1-1) will be called a *line*. The following theorem shows that P' is a model of the projective plane.

Theorem 3-1. *There is a one-to-one mapping σ: $P' \to P^2$ onto P^2 under which the lines of P' are carried in a one-to-one manner onto the lines of P^2. The mapping σ therefore preserves the incidence relations among lines and points.*

Proof. Let a_1 and a_2 be two numbers, not both zero. Let l be the line in E^2 that has the equation

$$-a_2X_1 + a_1X_2 = 0$$

Then the ideal point which consists of all lines parallel to l will be denoted by $(a_1|a_2)$. It follows that all symbols of the form $(ca_1|ca_2)$, with $c \neq 0$, denote this same ideal point. In terms of this notation, we define a mapping $\hat{\sigma}: E^* \to P^2$ as follows:

$$\hat{\sigma}(x_1, x_2, x_3) = \begin{cases} \left(\dfrac{x_1}{x_3}, \dfrac{x_2}{x_3}\right) & \text{if } x_3 \neq 0 \\ (x_1|x_2) & \text{if } x_3 = 0 \end{cases}$$

Since

$$\hat{\sigma}(cx_1, cx_2, cx_3) = \hat{\sigma}(x_1, x_2, x_3)$$

where $c \neq 0$, the mapping $\hat{\sigma}$ induces a mapping $\sigma: P' \to P^2$ given by the rule: If p' is a point of P' with the coordinates (x_1, x_2, x_3), then

$$\sigma(p') = \hat{\sigma}(x_1, x_2, x_3)$$

The reader may easily verify that σ is a one-to-one mapping of P' onto P^2.

Let a_1, a_2, a_3 be numbers which are not all zero. Let l' be the line in P' whose points satisfy the relation

$$a_1X_1 + a_2X_2 + a_3X_3 = 0$$

We suppose first that a_1 and a_2 are not both zero. Then (x_1, x_2, x_3) are coordinates of a point on l' if and only if

$$a_1x_1 + a_2x_2 + a_3x_3 = 0$$

and this relation holds if and only if one of the following two conditions hold: Either

$$a_1\left(\frac{x_1}{x_3}\right) + a_2\left(\frac{x_2}{x_3}\right) + a_3 = 0$$

or

$$a_1x_1 + a_2x_2 = 0$$

That is to say, (x_1, x_2, x_3) are coordinates of a point on l' if and only if $\hat{\sigma}(x_1, x_2, x_3)$ is on the line

$$a_1X_1 + a_2X_2 + a_3 = 0$$

in E^2 or is its ideal point.

We now suppose that $a_1 = a_2 = 0$. Then the points of P' which satisfy the relation $a_3X_3 = 0$ are precisely those which have coordinates of the form $(x_1, x_2, 0)$. Since $\hat{\sigma}(x_1, x_2, 0) = (x_1|x_2)$, we see that a point in P' has coordinates $(x_1, x_2, 0)$ if and only if its image under σ is on the ideal line. We have therefore proved that the mapping σ induces a mapping of the lines of P' onto the lines of P^2. And it is clear that this mapping is one-to-one.

We are now justified in using the numerical model P' in place of P^2 whenever it is convenient. Consequently we shall no longer distinguish between P' and P^2.

EXERCISES

1. Find the equation (in E^2) of the line through the points $(-1, 1)$ and $(-2|1)$.

2. Prove that the points (x_1, x_2, x_3), (y_1, y_2, y_3), and (z_1, z_2, z_3) in P' are collinear if and only if the rank of the matrix

$$\begin{pmatrix} x_1 & x_2 & x_3 \\ y_1 & y_2 & y_3 \\ z_1 & z_2 & z_3 \end{pmatrix}$$

is less than 3.

3. Prove that the lines

$$\begin{aligned} a_1X_1 + a_2X_2 + a_3X_3 &= 0 \\ b_1X_1 + b_2X_2 + b_3X_3 &= 0 \\ c_1X_1 + c_2X_2 + c_3X_3 &= 0 \end{aligned}$$

in P' have a common point if and only if the rank of the matrix

$$\begin{pmatrix} a_1 & a_2 & a_3 \\ b_1 & b_2 & b_3 \\ c_1 & c_2 & c_3 \end{pmatrix}$$

is less than 3.

3-2 THE DIFFERENTIABLE STRUCTURE OF P^2

Let $GL(3)$ be the general linear group on V^3. In Section 1-5 we have seen that we may look upon $GL(3)$ as the group of affinities of A^3 that keep a point fixed. We may assume that the fixed point is $(0, 0, 0)$. Under these circumstances, we have arranged matters so that the elements of $GL(3)$ act on E^3 in such a way as to carry straight lines through the origin onto straight lines through the origin.

Let ρ be an element of $GL(3)$. Then ρ induces a one-to-one mapping of P^2 onto itself since the points of P^2 are in one-to-one correspondence with the lines in E^3 which pass through the origin. We shall also denote this mapping of P^2 by ρ.

Theorem 3-2. *The elements of $GL(3)$ act as projective transformations of P^2 and, conversely, each projective transformation of P^2 may be effected by an element of $GL(3)$.*

Proof. If $\rho \in GL(3)$, then ρ carries any plane in E^3 which passes through the origin onto another plane through the origin. Consequently it carries a line in P^2 onto a line in P^2 and is therefore a projective transformation.

Conversely, let ρ be a projective transformation of P^2. Let l be the ideal line in P^2 and $l' = \rho(l)$. Since l and l' are associated with planes in E^3 which pass through the origin, it is clear that there is an element ρ_1 of $GL(3)$ such that $l' = \rho_1(l)$. Then $\rho_2 = \rho_1^{-1}\circ\rho$ is a projective transformation of P^2 which carries the ideal line onto itself. It follows that ρ_2 is an affinity of A^2 (the complement of the ideal line) and may be described by

$$(x_1, x_2, x_3) \rightarrow (y_1, y_2, y_3)$$

where

$$\frac{y_1}{y_3} = a_{11}\frac{x_1}{x_3} + a_{12}\frac{x_2}{x_3} + a_{13}$$

$$\frac{y_2}{y_3} = a_{21}\frac{x_1}{x_3} + a_{22}\frac{x_2}{x_3} + a_{23}$$

and the matrix

$$\begin{pmatrix} a_{11} & a_{12} \\ a_{21} & a_{22} \end{pmatrix}$$

is nonsingular. If we take $y_3 = x_3$, then ρ_2 is described by an element of $GL(3)$; hence so is $\rho = \rho_1\circ\rho_2$.

Corollary. *If l and m are any two lines (p and q are any two points) of P^2, then there is a projective transformation which carries l onto m (p onto q).*

Theorem 3-3. *There is one and only one topological structure on P^2 which satisfies the following conditions:*

(i) *The topology of P^2 induces on A^2 (the complement of the ideal line) its usual topology.*

(ii) *Every projective transformation of P^2 is a homeomorphism.*

Proof. Let E^* be the set of points (x_1, x_2, x_3) of E^3 excepting (0, 0, 0). Let σ be the mapping of E^* onto P^2 which carries (x_1, x_2, x_3) onto the point with these coordinates. We introduce into P^2 the natural topology that is induced on it from E^* by the mapping σ. We shall prove that this topology satisfies conditions i and ii.

Let S^* be the subset of E^* which consists of those points (x_1, x_2, x_3), with $x_3 \neq 0$. Define $\tau: S^* \rightarrow A^2$ by

$$\tau(x_1, x_2, x_3) = \left(\frac{x_1}{x_3}, \frac{x_2}{x_3}\right)$$

(see Figure 3-2). Let $S = \sigma(S^*)$. Let $\bar{\tau}: S \rightarrow A^2$ be the one-to-one mapping of S onto A^2 for which $\tau = \bar{\tau}\circ\sigma$. Condition i will be proved if we show that $\bar{\tau}$ is a homeomorphism.

Let U be an open subset of S. Then $\sigma^{-1}(U)$ is an open subset of S^*. Thus $\bar{\tau}(U) = \tau\circ\sigma^{-1}(U)$ is the image of an open set under the mapping τ.

It is not difficult to see that τ maps open sets onto open sets, and so $\bar{\tau}(U)$ is open. This proves that $\bar{\tau}^{-1}$ is continuous. On the other hand, if V is an open subset of A^2, then $\tau^{-1}(V)$ is an open subset of S^*. Since $\sigma^{-1}(\bar{\tau}^{-1}(V)) = \tau^{-1}(V)$, $\bar{\tau}^{-1}(V)$ is an open subset of S. This proves that $\bar{\tau}$ is continuous and completes the proof of condition i.

Any projective transformation ρ may be effected by an element of $GL(3)$, which certainly produces a homeomorphism of E^*. Since σ is an open mapping, it follows that ρ is a homeomorphism of P^2, and so condition ii is true.

To show that the topology on P^2 is uniquely determined by conditions i and ii, it is sufficient to prove that each point of P^2 is contained

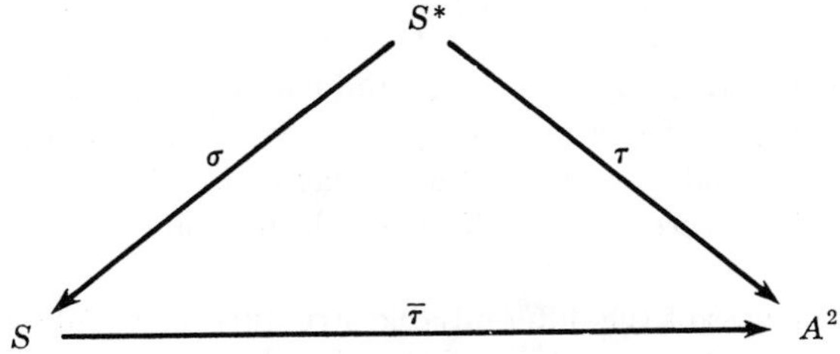

FIG. 3-2

in a set $\rho(S)$, where ρ is a suitable projective transformation. But this statement follows immediately from the corollary to Theorem 3-2.

Theorem 3-4. *There is one and only one differentiable structure on P^2 which satisfies the following conditions:*

(i) *The differentiable structure of P^2 induces on A^2 (by taking the coordinate neighborhoods of A^2 to be those of P^2 which are contained in A^2) its usual differentiable structure.*

(ii) *Every projective transformation of P^2 is a diffeomorphism.*

Proof. Let ρ_1 be the projective transformation of P^2 given by $\rho_1(x_1, x_2, x_3) = (x_2, x_3, x_1)$. Let $\rho_2 = \rho_1 \circ \rho_1$. Then every point of P^2 is in one of the three sets A^2, $\rho_1(A^2)$, or $\rho_2(A^2)$. We define $\varphi: A^2 \to E^2$ by

$$\varphi(x_1, x_2, x_3) = \left(\frac{x_1}{x_3}, \frac{x_2}{x_3}\right)$$

and $\varphi_1: \rho_1(A^2) \to E^2$ and $\varphi_2: \rho_2(A^2) \to E^2$ by $\varphi_1 = \varphi \circ \rho_1^{-1}$ and $\varphi_2 = \varphi \circ \rho_2^{-1}$, respectively. It is not difficult to prove that $\varphi_1 \circ \varphi^{-1}: \varphi(A_2 \cap \rho_1(A_2)) \to \varphi_1(A_2 \cap \rho_1(A_2))$ is a differentiable mapping, and similar statements hold for the mappings $\varphi_2 \circ \varphi_1^{-1}$ and $\varphi \circ \varphi_2^{-1}$ and the inverses of these three mappings. Consequently they define a differentiable structure on P^2 which satisfies condition i.

Let ρ be a projective transformation of P^2. In order to prove condition ii, it suffices to prove that ρ is differentiable, for a similar proof can be given for ρ^{-1}. Let (x_1, x_2, x_3) be a point in P^2. We shall assume that (x_1, x_2, x_3) is in A^2 and that $\rho(x_1, x_2, x_3)$ is in $\rho_1(A^2)$, but a similar argument can be given for any other pair of coordinate neighborhoods. If $(y_1, y_2, y_3) = \rho(x_1, x_2, x_3)$, then

$$\varphi_1 \circ \rho \circ \varphi^{-1}\left(\frac{x_1}{x_3}, \frac{x_2}{x_3}\right) = \varphi \circ \rho_1^{-1}(y_1, y_2, y_3) = \left(\frac{y_3}{y_2}, \frac{y_1}{y_2}\right)$$

Now the projective transformation ρ is given by

$$(y_1 \quad y_2 \quad y_3) = (u \quad v \quad 1)\,\mathfrak{M}$$

where $\mathfrak{M}$ is a matrix and $u = x_1/x_3$ and $v = x_2/x_3$ are the coordinates of points in A^2. Consequently y_3/y_2 and y_1/y_2 are differentiable functions of u and v, and it follows that $\varphi_1 \circ \rho \circ \varphi^{-1}$ is a differentiable mapping. Therefore ρ is a differentiable mapping, and condition ii is proved.

The uniqueness of the differentiable structure on P^2 follows immediately from conditions i and ii and the fact that every point of P^2 is in one of the sets A^2, $\rho_1(A^2)$, or $\rho_2(A^2)$.

EXERCISES

1. Prove that the only elements of $GL(3)$ which act as the identity transformation on P^2 are the nonzero numerical multiples of the identity element of $GL(3)$. Hence prove that two elements ρ, σ of $GL(3)$ produce the same projective transformation if and only if there is a nonzero number c such that $\sigma = c\rho$.

2. Let p_1, p_2, p_3, p_4 be points of P^2, no three of which are collinear. Prove that a projective transformation which keeps each of p_1, p_2, p_3, p_4 fixed must be the identity transformation. This means that a projective transformation ρ is completely determined by the images $\rho(p_1), \rho(p_2), \rho(p_3), \rho(p_4)$.

3. Let p_1, p_2, p_3, p_4 be points of P^2, no three of which are collinear. Let q_1, q_2, q_3, q_4 be another set of such points. Prove that there is a projective transformation ρ such that $\rho(p_1) = q_1$, $\rho(p_2) = q_2$, $\rho(p_3) = q_3$, $\rho(p_4) = q_4$.

4. Prove that P^2 is a compact topological space.

3-3 REAL AND COMPLEX PROJECTIVE SPACES OF HIGHER DIMENSIONS

Let E^* be the set of all points of E^{n+1} except $(0, \ldots, 0)$. Two points $(x_1, \ldots, x_{n+1})$ and $(y_1, \ldots, y_{n+1})$ are said to be equivalent if there is a number c such that $y_1 = cx_1, \ldots, y_{n+1} = cx_{n+1}$. This equivalence relation partitions E^* into mutually disjoint classes.

Let P^n denote the set of equivalence classes of E^*. P^n is called *projective n-space.*

Let σ be the identification mapping defined by

$$\sigma(x_1, \ldots, x_{n+1}) = \text{equivalence class containing } (x_1, \ldots, x_{n+1})$$

We shall give to P^n the topology that is induced on it from the topology in E^{n+1} by the mapping σ. For $n = 2$, P^2 is the projective plane which we have just studied. We shall leave to the reader the definition of a differentiable structure on P^n, which proceeds in a fashion completely analogous to the construction which we have carried out for P^2.

So far we have dealt exclusively with real euclidean and projective spaces. We wish to consider briefly now the analogous concepts when the real numbers are replaced by the complex numbers.

Let C^n consist of all n-tuples $(\alpha_1, \ldots, \alpha_n)$ where $\alpha_1, \ldots, \alpha_n$ are complex numbers. If we define the distance $d(\alpha, \beta)$ between two elements $\alpha = (\alpha_1, \ldots, \alpha_n)$ and $\beta = (\beta_1, \ldots, \beta_n)$ by

$$d(\alpha, \beta) = \sqrt{|\beta_1 - \alpha_1|^2 + \cdots + |\beta_n - \alpha_n|^2}$$

then C^n becomes a metric space with the same structure as E^{2n}. For if we put

$$\begin{aligned} \mathcal{R}\alpha_i &= \text{real part of } \alpha_i \\ \mathcal{I}\alpha_i &= \text{imaginary part of } \alpha_i \end{aligned}$$

we may identify the point $(\alpha_1, \ldots, \alpha_n)$ in C^n with the point

$$(\mathcal{R}\alpha_1, \mathcal{I}\alpha_1, \ldots, \mathcal{R}\alpha_n, \mathcal{I}\alpha_n)$$

in E^{2n}.

On the other hand, if α and β are elements of C^n and c is a complex number, we may define

$$\begin{aligned} \alpha + \beta &= (\alpha_1 + \beta_1, \ldots, \alpha_n + \beta_n) \\ c\alpha &= (c\alpha_1, \ldots, c\alpha_n) \end{aligned}$$

Then C^n becomes a vector space of dimension n over the complex numbers which we shall denote by $V^n{}_C$. The mapping

$$\alpha \to (\mathcal{R}\alpha_1, \mathcal{I}\alpha_1, \ldots, \mathcal{R}\alpha_n, \mathcal{I}\alpha_n)$$

is an isomorphism of $V^n{}_C$ onto V^{2n} *as real vector spaces.* The only way in which these two spaces differ on this level is that $V^n{}_C$ permits multiplication by complex numbers and V^{2n} does not. We shall assign to $V^n{}_C$ the same differentiable structure as that of V^{2n} if we merely wish to require that $V^n{}_C$ be a differentiable manifold. It is clear how we may identify C^n with $V^n{}_C$, so that C^n also has a differentiable structure.

For many purposes it is desirable to strengthen the differentiable structure of $V^n{}_C$ beyond that customary for real differentiable manifolds. This is done by restricting the differentiable functions on $V^n{}_C$ to be those complex-valued functions f for which

$$f(z_1, \ldots, z_n) = \text{convergent power series in } z_1, \ldots, z_n$$

in a neighborhood of each point. Such functions are said to be *complex-analytic*, and under such a restriction $V^n{}_C$ becomes a complex-analytic manifold.

From C^{n+1} we may construct complex-projective n-space. Let C^* be the set of all points of C^{n+1} except $(0, \ldots, 0)$. Two points α and β of C^* are called equivalent if there is a *complex number* c such that $\beta_1 = c\alpha_1, \ldots, \beta_{n+1} = c\alpha_{n+1}$. The set of equivalence classes of C^* will be denoted by $P^n{}_C$. The reader may introduce a topology and a differentiable structure into $P^n{}_C$ by analogy with the procedure for P^n.

We shall attempt to give some geometric insight into the structure of the real projective spaces. Consider first P^1. The equivalence classes of $E^* \subset E^2$ are the straight lines which pass through (0, 0), except for the point (0, 0). Each of these equivalence classes intersects the upper half of the unit circle,

$$x_1^2 + x_2^2 = 1 \qquad x_2 \geqq 0$$

in one point, except the one class that intersects it in the two end points (1, 0) and (−1, 0). If we identify these end points, the resulting space is topologically equivalent to a circle (1-sphere). It is not difficult to see that the one-to-one correspondence which we have just described between P^1 and the circle is a homeomorphism. Thus P^1 is topologically equivalent to a circle.

A similar construction is possible for P^2. Again the equivalence classes of $E^* \subset E^3$ are the lines through (0, 0, 0), except for the point (0, 0, 0). These lines intersect the upper half of the unit sphere,

$$x_1^2 + x_2^2 + x_3^2 = 1 \qquad x_3 \geqq 0$$

in one point, except the lines that lie in the x_1, x_2-plane. Each of the lines in the x_1, x_2-plane intersects the circular boundary,

$$x_1^2 + x_2^2 = 1 \qquad x_3 = 0$$

of the half sphere in two points which are diametrically opposite one another. If we identify these diametrically opposite points of the boundary of the half sphere, we obtain a space that is topologically equivalent to P^2.

Another, less concrete representation for P^1 is the following more symmetric construction. Each line in E^2 through (0, 0) intersects the unit circle,

$$x_1^2 + x_2^2 = 1$$

in two diametrically opposite points. The space obtained from the circle by identifying diametrically opposite points is therefore topologically equivalent to P^1. This is precisely the identification which we made on the boundary of the half sphere on obtaining P^2. Consequently we may obtain P^2 from the half sphere by identifying the circular boundary so as to make it topologically equivalent to P^1.

In general, to obtain a topological model for P^n we begin with the lines in E^{n+1} which pass through $(0, \ldots, 0)$. Each of these lines intersects the half unit n-sphere,

$$x_1^2 + \cdots + x_{n+1}^2 = 1 \qquad x_{n+1} \geqq 0$$

in one point, except the lines in the hyperplane $x_{n+1} = 0$. The boundary of the half n-sphere has the equation

$$x_1^2 + \cdots + x_n^2 = 1 \qquad x_{n+1} = 0$$

Hence it is the $(n - 1)$-sphere. In order to obtain P^n from the half n-sphere, we must identify the antipodal points of its boundary, since each line of E^{n+1} which passes through $(0, \ldots, 0)$ and which lies in the hyperplane $x_{n+1} = 0$ intersects the boundary $(n - 1)$-sphere in two antipodal points.

We may also obtain P^n from the full unit n-sphere by identifying antipodal points. For each line in E^{n+1} through $(0, \ldots, 0)$ intersects the unit n-sphere in precisely two points which are antipodal. But this is precisely the identification which we made in the boundary of the half n-sphere in our previous construction. Consequently P^n is obtained from the half n-sphere by identifying the boundary points in such a way that the boundary becomes P^{n-1}.

To see how the complex-projective spaces differ from the real-projective spaces we shall consider $P^1{}_C$. The other complex-projective spaces cannot be treated adequately without the notion of fiber bundle which we shall introduce in Chapter 9.

For every point in $P^1{}_C$ with coordinates (z_1, z_2) except one, the coordinate z_2 will be different from zero. Consequently, except for that one point, we may associate the complex number z_1/z_2 with the point whose coordinates are (z_1, z_2). This gives a one-to-one correspondence between the set P^* of points (z_1, z_2) of $P^1{}_C$ with $z_2 \neq 0$ and the complex plane, which is actually a homeomorphism. We then map the complex plane onto the unit 2-sphere, excepting the point (0, 0, 1), by a homeomorphism called stereographic projection which we shall now describe.

Let n be the point (0, 0, 1). Take the x_1, x_2-plane to be the complex plane, so that $z_1/z_2 = x_1 + ix_2$. Join n to the point $z_1/z_2 = (x_1, x_2, 0)$ by a straight line. This straight line intersects the unit sphere in one point p which is different from n. p is called the stereographic projection of z_1/z_2 onto the unit sphere (see Figure 3-3). For fixed z_1,

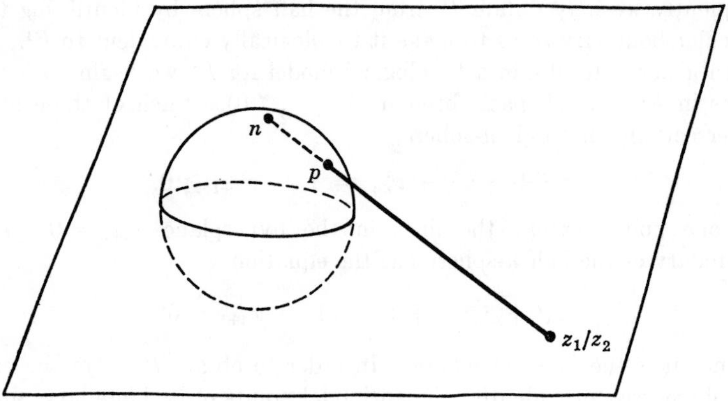

FIG. 3-3

as z_2 approaches 0 the stereographic projection p of z_1/z_2 approaches the point n. Consequently, we should map the point $(z, 0)$ onto n. In this way we produce a homeomorphism of $P^1{}_C$ onto the 2-sphere.

Theorem 3-5. *P^n and $P^n{}_C$ are compact topological spaces.*

Proof. P^n is obtained from the n-sphere by identifying antipodal points. Since the n-sphere is compact and the identification mapping is continuous, the image P^n must be compact.

$P^n{}_C$ is obtained from the compact subset of C^{n+1} defined by the equation

$$|z_1|^2 + \cdots + |z_{n+1}|^2 = 1$$

[this subset is the $(2n + 1)$-sphere in E^{2n+2}] by identifying suitable points. As before, the image must be compact.

EXERCISES

1. A hyperplane in P^n is the set of all points whose coordinates satisfy a relation of the form

$$a_1X_1 + \cdots + a_{n+1}X_{n+1} = 0$$

where $a_1, \ldots, a_{n+1}$ are not all zero. Prove that, if $(x_1, \ldots, x_{n+1}), \ldots, (w_1, \ldots, w_{n+1})$ are n points such that the matrix

$$\begin{pmatrix} x_1 & \cdots & x_{n+1} \\ \cdots & \cdots & \cdots \\ w_1 & \cdots & w_{n+1} \end{pmatrix}$$

has rank n, there is one and only one hyperplane which contains these points.

2. Prove that, if $a_1X_1 + \cdots + a_{n+1}X_{n+1} = 0, \ldots, g_1X_1 + \cdots + g_{n+1}X_{n+1} = 0$ are n hyperplanes such that the matrix

$$\begin{pmatrix} a_1 & \cdots & a_{n+1} \\ \cdots & \cdots & \cdots \\ g_1 & \cdots & g_{n+1} \end{pmatrix}$$

has rank n, these hyperplanes have one and only one point in common.

3. A projective transformation of P^n is a one-to-one mapping of P^n onto itself that carries lines onto lines. Prove that a one-to-one mapping of P^n onto itself is a projective transformation if and only if it carries hyperplanes onto hyperplanes.

4. Prove that every projective transformation of P^n can be effected by an element of $GL(n + 1)$.

5. State and prove the analogues for P^n of Theorems 3-3 and 3-4.

6. Give an example of a differentiable function on $V^n{}_C$ which is not complex-analytic.

7. Show that the homeomorphism $(z_1, z_2) \to (x_1, x_2, x_3)$ of $P^1{}_C$ onto the 2-sphere can be described analytically by

$$x_1 = \frac{2\Re z_1\bar{z}_2}{z_1{}^2 + z_2{}^2} \qquad x_2 = \frac{2\Im z_1\bar{z}_2}{z_1{}^2 + z_2{}^2} \qquad x_3 = \frac{z_1{}^2 - z_2{}^2}{z_1{}^2 + z_2{}^2}$$

where $\bar{z}_2$ is the conjugate complex number to z_2.

8. Prove that there is a one-to-one regular differentiable mapping of P^n into $P^n{}_C$.

3-4 ALGEBRAIC VARIETIES

If f is a polynomial in the variables $x_1, \ldots, x_n$, then f is a function which is defined on all of E^n. Let $f_1, \ldots, f_r$ be polynomials on E^n. Then the set of points $(x_1, \ldots, x_n)$ such that

$$\begin{gathered} f_1(x_1, \ldots, x_n) = 0 \\ \cdots\cdots\cdots \\ f_r(x_1, \ldots, x_n) = 0 \end{gathered}$$

is called an *affine algebraic variety*. An affine variety need not be a differentiable variety in E^n. In Section 2-1 we have seen the conditions under which it is a differentiable variety.

The close relationship between E^{n+1} and P^n enables us to define algebraic varieties in P^n, in spite of the fact that one cannot define polynomial functions on P^n.

A polynomial function f on E^{n+1} is called homogeneous if it satisfies

$$f(tx_1, \ldots, tx_{n+1}) = t^d f(x_1, \ldots, x_{n+1})$$

where d is the degree of f. Every term of f then has the degree d. It is clear that if $(a_1, \ldots, a_{n+1})$ are the coordinates of a point in P^n and if $f\ a_1, \ldots, a_{n+1}) = 0$, then f is also zero for all other coordinates of that same point. Let $h_1, \ldots, h_r$ be homogeneous polynomials on E^{n+1}. The points of P^n whose coordinates $(x_1, \ldots, x_{n+1})$ satisfy

$$\begin{gathered} h_1(x_1, \ldots, x_{n+1}) = 0 \\ \cdots\cdots\cdots\cdots \\ h_r(x_1, \ldots, x_{n+1}) = 0 \end{gathered}$$

will be called a *projective algebraic variety*.

Theorem 3-6. *Every projective algebraic variety is a compact topological space.*

Proof. We have seen in Theorem 3-5 that P^n is a compact space. Therefore it is sufficient to demonstrate that a projective variety in P^n is a closed subset of P^n. Let the homogeneous polynomials $h_1, \ldots, h_r$ define a projective variety V, and let V_j be the variety that is defined by the single polynomial h_j. Then

$$V = V_1 \cap \cdots \cap V_r$$

Thus it suffices to show that each of $V_1, \ldots, V_r$ is a closed set.

Let $h(y_1, \ldots, y_{n+1})$ be a homogeneous polynomial. The set S of points $(y_1, \ldots, y_{n+1})$ in E^{n+1} for which $h(y_1, \ldots, y_{n+1}) = 0$ is closed, since the mapping $h: E^{n+1} \to E^1$ is continuous and $S = h^{-1}(\{0\})$. In accordance with the way in which we define the topology of P^n from the topology in E^{n+1}, the projective variety defined by h must be a closed subset of P^n, which completes the proof.

Since we may regard E^n as a subspace (or even submanifold) of P^n, we may imbed any affine algebraic variety V of E^n in the projective space P^n. We shall see that it is possible to complete V to a projective algebraic variety by the adjunction of suitable points at infinity. The procedure also provides us with a "compactification" of the locally compact space V, since the completed variety is a compact space, according to Theorem 3-6. We shall illustrate the procedure with an algebraic curve in the plane. The reader may easily generalize it to an affine variety in E^n.

Let $f(x_1, x_2)$ be a polynomial function on E^2 which defines an affine algebraic variety V^*. A point (x_1, x_2) is in V^* if and only if $f(x_1, x_2) = 0$.

We may identify the point (x_1, x_2) in E^2 with the point in P^2 that has the coordinates $(x_1, x_2, 1)$. Then V^* becomes the subset of P^2 consisting of those points $(x_1, x_2, 1)$ of P^2 for which $f(x_1, x_2) = 0$. More generally, we can say that the point (y_1, y_2, y_3) is in V^* if and only if $y_3 \neq 0$ and $f(y_1/y_3, y_2/y_3) = 0$. Let d be the degree of f. Then the polynomial

$$h(y_1, y_2, y_3) = y_3{}^d f\left(\frac{y_1}{y_3}, \frac{y_2}{y_3}\right)$$

is homogeneous. Furthermore, (y_1, y_2, y_3) is a point in V^* if and only if $y_3 \neq 0$ and $h(y_1, y_2, y_3) = 0$. Thus we see that V^* is a subset of the projective algebraic variety V in P^2 which is defined by the polynomial h. In fact, V^* is that subset of V which is obtained by removing the points at infinity (those with $y_3 = 0$).

Of course, E^2 may be imbedded in P^2 in an infinite number of ways, depending upon which line in P^2 is chosen to be the line at infinity. Hence it appears that V^* may be completed in many different ways. Actually, for any two completions of V^* there is a projective transformation of P^2 which carries the one on the other, so that they are equivalent as far as the geometry of the projective plane is concerned.

Let V be a projective algebraic variety in P^n defined by the homogeneous polynomials $h_1, \ldots, h_r$ of degrees $d_1, \ldots, d_r$, respectively. Consider the jacobian matrix

$$\mathcal{J}(y_1, \ldots, y_{n+1}) = \begin{pmatrix} \dfrac{\partial h_1}{\partial y_1} & \cdots & \dfrac{\partial h_1}{\partial y_{n+1}} \\ \cdots & \cdots & \cdots \\ \dfrac{\partial h_r}{\partial y_1} & \cdots & \dfrac{\partial h_r}{\partial y_{n+1}} \end{pmatrix}$$

where all the partial derivatives are evaluated at the same point $(y_1, \ldots, y_{n+1})$ of E^{n+1}. Since

$$\mathcal{J}(cy_1, \ldots, cy_{n+1}) = \mathcal{C}\mathcal{J}(y_1, \ldots, y_{n+1})$$

where
$$\mathcal{C} = \begin{pmatrix} c^{d_1-1} & & 0 \\ & \ddots & \\ 0 & & c^{d_r-1} \end{pmatrix}$$

it follows that the rank of $\mathcal{J}(y_1, \ldots, y_{n+1})$ depends only upon the point of P^n whose coordinates are $(y_1, \ldots, y_{n+1})$ and not on the actual coordinates used. We shall say that V is *nonsingular* if the rank of $\mathcal{J}(y_1, \ldots, y_{n+1})$ is the same at all points of V.

Let $\rho: P^n \to P^n$ be a projective transformation. Theorem 3-2 may be generalized to P^n, and from this it follows that the relationship between the coordinates $(y_1, \ldots, y_{n+1})$ of a point in P^n and the coordinates $(y'_1, \ldots, y'_{n+1})$ of its image under ρ is given by

$$(y'_1 \cdots y'_{n+1}) = (y_1 \cdots y_{n+1})\, \mathfrak{M}$$

where $\mathfrak{M}$ is a nonsingular matrix. If V is a projective variety defined by the homogeneous polynomials $h_1, \ldots, h_r$, then $\rho(V)$ is the projective variety defined by the homogeneous polynomials $g_1, \ldots, g_r$, where

$$\begin{aligned} g_1(y'_1, \ldots, y'_{n+1}) &= h_1(y_1, \ldots, y_{n+1}) \\ &\cdots\cdots\cdots\cdots \\ g_r(y'_1, \ldots, y'_{n+1}) &= h_r(y_1, \ldots, y_{n+1}) \end{aligned}$$

Let

$$\mathcal{J}'(y'_1, \ldots, y'_{n+1}) = \begin{pmatrix} \dfrac{\partial g_1}{\partial y'_1} & \cdots & \dfrac{\partial g_1}{\partial y'_{n+1}} \\ \cdots & \cdots & \cdots \\ \dfrac{\partial g_r}{\partial y'_1} & \cdots & \dfrac{\partial g_r}{\partial y'_{n+1}} \end{pmatrix}$$

Since

$$\frac{\partial g_j}{\partial y'_i} = \sum_{k=1}^{n+1} \frac{\partial h_j}{\partial y_k}\frac{\partial y_k}{\partial y'_i} = \sum_{k=1}^{n+1} c_{ik}\frac{\partial h_j}{\partial y_k}$$

where $\mathfrak{M}^{-1} = (c_{ij})$, it follows that

$$\mathcal{J}'(y'_1, \ldots, y'_{n+1}) = \mathcal{J}(y_1, \ldots, y_{n+1})\, \mathfrak{N}$$

where $\mathfrak{N}$ is the transpose of $\mathfrak{M}^{-1}$. We may conclude that the matrices $\mathcal{J}'(y'_1, \ldots, y'_{n+1})$ and $\mathcal{J}(y_1, \ldots, y_{n+1})$ have the same rank at corresponding points under the transformation ρ. Consequently $\rho(V)$ is nonsingular if and only if V is nonsingular. In other words, *the nonsingularity of a projective variety is a projective invariant.*

Theorem 3-7. *A nonsingular projective algebraic variety can be given the structure of a differentiable manifold.*

Proof. Let V be a projective algebraic variety, and let p be a point of V. By means of a projective transformation ρ, we may carry p onto the point with coordinates $(0, \ldots, 0, 1)$. We identify the point $(x_1, \ldots, x_n)$ of E^n with the point of P^n whose coordinates are $(x_1, \ldots, x_n, 1)$. Then $\rho(p)$ is in E^n.

Let $h_1, \ldots, h_r$ be the homogeneous polynomials which define $\rho(V)$. Let

$$f_1(x_1, \ldots, x_n) = h_1(x_1, \ldots, x_n, 1)$$
$$\cdots\cdots\cdots\cdots\cdots\cdots$$
$$f_r(x_1, \ldots, x_n) = h_r(x_1, \ldots, x_n, 1)$$

Then $\rho(V) \cap E^n$ is the affine algebraic variety defined by the polynomials $f_1, \ldots, f_r$. We shall show that $\rho(V) \cap E^n$ satisfies the conditions described in Section 2-1 for a differentiable variety in E^n.

Let $\mathcal{J}(x_1, \ldots, x_n)$ be the $r \times n$ matrix

$$\mathcal{J}(x_1, \ldots, x_n) = \begin{pmatrix} \dfrac{\partial f_1}{\partial x_1} & \cdots & \dfrac{\partial f_1}{\partial x_n} \\ \cdots & \cdots & \cdots \\ \dfrac{\partial f_r}{\partial x_1} & \cdots & \dfrac{\partial f_r}{\partial x_n} \end{pmatrix}$$

where the derivatives are evaluated at $(x_1, \ldots, x_n)$. In terms of the polynomials $h_1(y_1, \ldots, y_{n+1}), \ldots, h_r(y_1, \ldots, y_{n+1})$, we have

$$\frac{\partial f_i}{\partial x_j} = \left.\frac{\partial h_i}{\partial y_j}\right|_{(x_1, \ldots, x_n, 1)}$$

for $i = 1, \ldots, r$ and $j = 1, \ldots, n$. Thus $\mathcal{J}(x_1, \ldots, x_n)$ is the same as the jacobian matrix of $\rho(V)$ at the point $(x_1, \ldots, x_n, 1)$, except that the last column is missing. We shall show that this missing column is a linear combination of the others, and consequently the rank of $\mathcal{J}(x_1, \ldots, x_n)$ is equal to the rank of the jacobian matrix of $\rho(V)$. Since $\rho(V)$ is nonsingular, it follows that $\mathcal{J}(x_1, \ldots, x_n)$ has the same rank at all points of $\rho(V) \cap E^n$.

Consider the identity

$$h_i(ty_1, \ldots, ty_{n+1}) = t^d h_i(y_1, \ldots, y_{n+1})$$

where d is the degree of h_i. If we differentiate this identity with respect to t and set t equal to 1, then

$$y_1 \frac{\partial h_i}{\partial y_1} + \cdots + y_{n+1} \frac{\partial h_i}{\partial y_{n+1}} = d\, h_i(y_1, \ldots, y_{n+1})$$

At the points of $\rho(V) \cap E^n$ we have $h_i(x_1, \ldots, x_n, 1) = 0$, and so

$$\left.\frac{\partial h_i}{\partial y_{n+1}}\right|_{(x_1, \ldots, x_n, 1)} = -\sum_{j=1}^{n} x_j \left.\frac{\partial h_i}{\partial y_j}\right|_{(x_1, \ldots, x_n, 1)}$$

for $i = 1, \ldots, r$. This proves the assertion.

It follows that $\rho(V) \cap E^n$ is the differentiable variety in E^n defined by the equations

$$f_1(x_1, \ldots, x_n) = 0$$
$$\cdots\cdots\cdots\cdots$$
$$f_r(x_1, \ldots, x_n) = 0$$

We shall adopt as the differentiable structure of V in the neighborhood of p the differentiable structure of $\rho(V) \cap E^n$ transferred to p by the homeomorphism ρ^{-1}. Let U be a coordinate neighborhood of $(0, \ldots, 0, 1)$ in the manifold $\rho(V) \cap E^n$. Let $\varphi: U \to E^m$ be the coordinate mapping. Then $\varphi \circ \rho: \rho^{-1}(U) \to E^m$ is a homeomorphism onto an open set in E^m. The space V is covered by the sets of the form $\rho^{-1}(U)$ as we allow the point p to range over V. We shall show that these sets serve as coordinate neighborhoods of V.

Let $\rho': P^n \to P^n$ be another projective transformation. Let U' be a coordinate neighborhood of $(0, \ldots, 0, 1)$ in the differentiable variety $\rho'(V) \cap E^n$. Let $\varphi': U' \to E^m$ be the coordinate mapping. We suppose that $\rho^{-1}(U) \cap \rho'^{-1}(U')$ is not empty. It is then our task to prove that the mapping $\varphi' \circ \rho' \circ (\varphi \circ \rho)^{-1}$ is differentiable where it is defined. But this mapping is the composition of the mappings φ^{-1}, $\rho' \circ \rho^{-1}$, and φ'. Consequently it suffices to prove that $\rho' \circ \rho^{-1}$ is differentiable as a mapping of the subset U of the manifold $\rho(V) \cap E^n$ into the manifold $\rho'(V) \cap E^n$. In accordance with the remarks following Theorem 2-1, this is equivalent to proving that $\rho' \circ \rho^{-1}$ is a differentiable mapping of U into E^n. Moreover, by Theorem 2-1, a function on U is differentiable at a point of U if it is the restriction of a differentiable function on an open set in E^n. This means that $\rho' \circ \rho^{-1}$ is differentiable at a point of U if it is the restriction of a differentiable mapping on some open set in E^n. But this last statement is clearly true since $\rho' \circ \rho^{-1}$ is a projective transformation when applied to all of P^n.

EXERCISES

1. Prove that the projective variety defined by $y_1^2 - y_2^2 - y_3^2 = 0$ in P^2 is nonsingular.

2. Prove that the projective variety defined by

$$y_1 y_2 y_3 - y_4^3 = 0$$
$$y_1^2 + y_2^2 + y_3^2 - y_4^2 = 0$$

in P^3 is nonsingular.

3. Show that the completions of the circle $x_1^2 + x_2^2 = 1$ and the hyperbola $x_1^2 - x_2^2 = 1$ in P^2 are projectively equivalent, in the sense that there is a projective transformation which carries one onto the other. Do the same for the circle $x_1^2 + x_2^2 = 1$ and the parabola $x_1^2 = x_2$.

CHAPTER 4

The Tangent Bundle of a Differentiable Manifold

4-1 VECTOR FIELDS ON E^n

Let U be an open subset of E^n. A *vector field* on U is a correspondence χ which assigns to each point x in U a vector $\chi(x)$ of the tangent space $T(x)$. The tangent vectors $(\partial/\partial x_1)_x, \ldots, (\partial/\partial x_n)_x$ form a basis of $T(x)$, and so we may write

$$\chi(x) = a_1(x)\frac{\partial}{\partial x_1} + \cdots + a_n(x)\frac{\partial}{\partial x_n}$$

Since the coefficients $a_1, \ldots, a_n$ may change from one point x to another, they will be regarded as functions defined on U.

If f is any differentiable function on U, we may form the "product"

$$\langle \chi(x), df\rangle = \sum_{i=1}^{n} a_i(x)\frac{\partial f}{\partial x_i} \tag{4-1-1}$$

at each point of U. Then $\langle \chi(x), df\rangle$ is also a function on U. If for every differentiable function f on U the corresponding function $\langle \chi, df\rangle$ is also differentiable, the vector field χ is said to be *differentiable* on U. It is clear from relation (4-1-1) that, if the functions $a_1, \ldots, a_n$ are differentiable on U, then χ is differentiable on U. Conversely, if χ is differentiable on U, we may take f to be the function that assigns to the point x its ith coordinate x_i. Then $\langle \chi, df\rangle = a_i$, and so $a_1, \ldots, a_n$ are differentiable functions. Thus the vector field χ is differentiable if and only if all the coefficient functions $a_1(x), \ldots, a_n(x)$ are differentiable on U. In particular, the vector field χ for which $\chi(x) = (\partial/\partial x_i)_x$ for all x is differentiable. It will be denoted by $\partial/\partial x_i$.

4-2 THE TANGENT BUNDLE

In this section we shall see that we may look upon a differentiable vector field as a differentiable mapping.

Let U be an open subset of E^n. Let $\mathfrak{T}(U)$ consist of all ordered pairs (x, α), where $x \in U$ and $\alpha \in T(x)$. If χ is a vector field on U, we may associate with χ the mapping $x \rightarrow (x, \chi(x))$ of U into $\mathfrak{T}(U)$. We shall also denote this mapping by χ. Thus $\chi(x)$ will denote either a vector from $T(x)$ or a pair (x, α) from the set $\mathfrak{T}(U)$, depending upon the context.

We shall show that it is possible to give a differentiable structure to $\mathfrak{T}(U)$ in such a way that the differentiable vector fields on U are precisely those mappings of the form $x \rightarrow (x, \alpha)$ which are differentiable. Let (x, α) be a point of $\mathfrak{T}(U)$. Then

$$\alpha = \sum_{i=1}^{n} c_i \left(\frac{\partial}{\partial x_i}\right)_x$$

The mapping

$$(x, \alpha) \rightarrow (x_1, \ldots, x_n, c_1, \ldots, c_n)$$

of $\mathfrak{T}(U)$ onto the subset $U \times E^n$ of E^{2n} is one-to-one. We may give to $\mathfrak{T}(U)$ the topology that makes this mapping a homeomorphism, and then we may regard the mapping as a coordinate mapping. In this way $\mathfrak{T}(U)$ is entirely covered with one coordinate neighborhood and becomes a differentiable manifold of dimension $2n$. The manifold $\mathfrak{T}(U)$ is called the *tangent bundle* over U.

Let $\chi = a_1(x)(\partial/\partial x_1) + \cdots + a_n(x)(\partial/\partial x_n)$ be a vector field on U. As the mapping

$$x \rightarrow (x_1, \ldots, x_n, a_1(x), \ldots, a_n(x))$$

into $\mathfrak{T}(U)$, χ is differentiable if and only if the functions $a_1, \ldots, a_n$ are differentiable, that is, if and only if χ is a differentiable vector field. Let $\pi\colon \mathfrak{T}(U) \rightarrow U$ be defined by $\pi(x, \alpha) = x$. From the point of view of the tangent bundle, a differentiable vector field on U is a differentiable mapping $\chi\colon U \rightarrow \mathfrak{T}(U)$ such that $\pi \circ \chi$ is the identity mapping. The mapping π is clearly differentiable. If we introduce the corresponding linear mappings which are induced on the tangent spaces, we obtain $\pi_* \circ \chi_* = 1$. Consequently χ is regular at each point of U.

EXERCISE

Let $\chi: U \to \mathfrak{T}(U)$ be any mapping such that $\pi \circ \chi$ is the identity mapping. Prove that χ is continuous if and only if for every differentiable function f on U the function $\langle \chi, df \rangle$ is continuous on U.

4-3 ELEMENTARY PROPERTIES OF VECTOR FIELDS

Let U be an open subset of E^n. Let $\mathfrak{D}$ be the ring of all functions that are differentiable on U. Let $\mathfrak{X}$ be the set of all differentiable vector fields on U. For the purposes of this section, we shall not need the tangent bundle $\mathfrak{T}(U)$, and so we shall take the point of view of Section 4-1. Then $\mathfrak{X}$ may be regarded as a module over $\mathfrak{D}$ in the following way: Let χ_1 and χ_2 be elements of $\mathfrak{X}$, and let f be an element of $\mathfrak{D}$. Define

$$(\chi_1 + \chi_2)(x) = \chi_1(x) + \chi_2(x)$$
$$(f\chi_1)(x) = f(x)\chi_1(x)$$

for all x in U. These relations define vector fields $\chi_1 + \chi_2$ and $f\chi_1$ on U which are clearly differentiable. The resulting operations make $\mathfrak{X}$ a module over $\mathfrak{D}$.

The vector fields $\partial/\partial x_1, \ldots, \partial/\partial x_n$ span $\mathfrak{X}$ over $\mathfrak{D}$, and they are linearly independent, in the sense that if $\Sigma_{i=1}^n f_i(\partial/\partial x_i) = 0$ the functions $f_1, \ldots, f_n$ are all zero. When we speak of the linear independence of vector fields on U, we shall always mean linear independence with respect to $\mathfrak{D}$.

Let $\delta: U \to V$ be a one-to-one differentiable mapping of U onto an open set V in E^n such that δ^{-1} is also differentiable. Then δ induces a linear transformation $\delta_*: T(x) \to T(\delta(x))$ for each x in U. Let χ be a vector field on U. If y is a point of V, we may define

$$\chi^\delta(y) = \delta_*[\chi(\delta^{-1}(y))]$$

Then χ^δ is a vector field on V which is said to be induced by δ from χ. If g is any function which is differentiable on V, then $\langle \chi^\delta, g \rangle = \langle \chi, g \circ \delta \rangle$ (see Section 1-3). Hence if χ is differentiable on U, then χ^δ is differentiable on V.

Let $(y_1, \ldots, y_n) = \delta(x_1, \ldots, x_n)$, and let

$$\chi(x) = \sum_{j=1}^n a_j(x) \left(\frac{\partial}{\partial x_j}\right)_x$$

We know that there are functions $b_1, \ldots, b_n$ on V such that

$$\chi^\delta(y) = \sum_{i=1}^{n} b_i(y)\left(\frac{\partial}{\partial y_i}\right)_y$$

It is natural to ask how the functions $b_1, \ldots, b_n$ are related to the functions $a_1, \ldots, a_n$. Since

$$\delta_*[\chi(x)] = \sum_{j=1}^{n} a_j(x)\,\delta_*\left[\left(\frac{\partial}{\partial x_j}\right)_x\right]$$

and

$$\delta_*\left[\left(\frac{\partial}{\partial x_j}\right)_x\right] = \sum_{i=1}^{n} \frac{\partial y_i}{\partial x_j}\left(\frac{\partial}{\partial y_i}\right)_{\delta(x)}$$

it follows that

$$b_i(y) = \sum_{j=1}^{n} \frac{\partial y_i}{\partial x_j}\, a_j(\delta^{-1}(y)) \tag{4-3-1}$$

The collection of n-tuples of functions $\{a_1, \ldots, a_n\}$, one collection for each coordinate neighborhood and all related by (4-3-1), is called a contravariant vector in the classical literature of differential geometry. The relation (4-3-1) is called the transformation law for a contravariant vector.

EXERCISES

1. Let

$$\chi_1(x_1, x_2) = (x_1 + x_2)\frac{\partial}{\partial x_1} - \frac{\partial}{\partial x_2}$$

$$\chi_2(x_1, x_2) = (x_2^2 + 1)\frac{\partial}{\partial x_1} + x_1\frac{\partial}{\partial x_2}$$

for all points (x_1, x_2) in E^2. Prove that χ_1 and χ_2 are linearly independent differentiable vector fields on E^2. Let

$$\chi(x_1, x_2) = (x_1^2 + x_2^2)\frac{\partial}{\partial x_1} + (x_1^2 - x_2^2)\frac{\partial}{\partial x_2}$$

for (x_1, x_2) in E^2. Find functions b_1 and b_2 such that $\chi = b_1\chi_1 + b_2\chi_2$.

2. Let δ: $E^2 \to E^2$ be the diffeomorphism of Exercise 7 in Section 1-3, and let χ_1, χ_2, and χ be the vector fields of the preceding exercise. Find the vector fields χ_1^δ, χ_2^δ, χ^δ.

3. Let $\chi_1, \ldots, \chi_n$ be linearly independent differentiable vector fields on an open subset U of E^n. Let

$$\chi = a_1\chi_1 + \cdots + a_n\chi_n$$

Prove that χ is differentiable on U if and only if the functions $a_1, \ldots, a_n$ are differentiable on U.

4. Let $\chi_1, \ldots, \chi_n$ be the vector fields of the preceding exercise. Let δ be a diffeomorphism of E^n. Prove that the vector fields $\chi_1^\delta, \ldots, \chi_n^\delta$ are linearly independent on $\delta(U)$.

5. Let δ be the diffeomorphism of Exercise 8 in Section 1-3. Find

$$\left(\frac{\partial}{\partial x_1}\right)^\delta \quad \left(\frac{\partial}{\partial x_2}\right)^\delta \quad \left(\frac{\partial}{\partial x_3}\right)^\delta$$

Let

$$\chi(x_1, x_2, x_3) = x_1x_2 \frac{\partial}{\partial x_1} + x_2x_3 \frac{\partial}{\partial x_2} + x_3x_1 \frac{\partial}{\partial x_3}$$

Find the functions b_1, b_2, b_3 such that

$$\chi = b_1 \left(\frac{\partial}{\partial x_1}\right)^\delta + b_2 \left(\frac{\partial}{\partial x_2}\right)^\delta + b_3 \left(\frac{\partial}{\partial x_3}\right)^\delta$$

4-4 FLOWS IN E^n

As we shall see, it is also possible to look upon a differentiable vector field as a system of differential equations. We shall first consider an example.

Let χ be the vector field

$$\chi = x_1 \frac{\partial}{\partial x_1} + x_2 \frac{\partial}{\partial x_2}$$

on E^2. We may attempt to visualize this vector field by drawing a directed line segment whose components are x_1, x_2 at the point (x_1, x_2). The resulting picture is given in Figure 4-1.

We shall undertake to solve the following geometric problem: Find a family $\mathfrak{F}$ of differentiable parametrized curves such that, if $\gamma \in \mathfrak{F}$ and γ passes through the point x, the tangent vector to γ at x is the given vector $\chi(x)$. In our example we see immediately that the straight lines through the origin will have tangent vectors with the correct directions. But these lines will have to be parametrized suitably if their tangent vectors are to be vectors of the given vector field χ.

Consider the straight line $x_2 = mx_1$. Let us suppose that it is possible to parametrize this straight line so that it has the correct tangent vectors. If this parametrization is given by $x_1 = f(t)$, we must have

$$f'(t) \frac{\partial}{\partial x_1} + mf'(t) \frac{\partial}{\partial x_2} = f(t) \frac{\partial}{\partial x_1} + mf(t) \frac{\partial}{\partial x_2}$$

Hence $f(t)$ must satisfy the differential equation $f'(t) = f(t)$. The general solution of this equation is $f(t) = ce^t$; clearly $\gamma(t) = (ce^t, mce^t)$ is a solution of our problem.

For the purposes of this section, our former definition of a differentiable parametrized curve in E^n and its tangent vector must be relaxed a bit, as is indicated by the example that we have just examined.

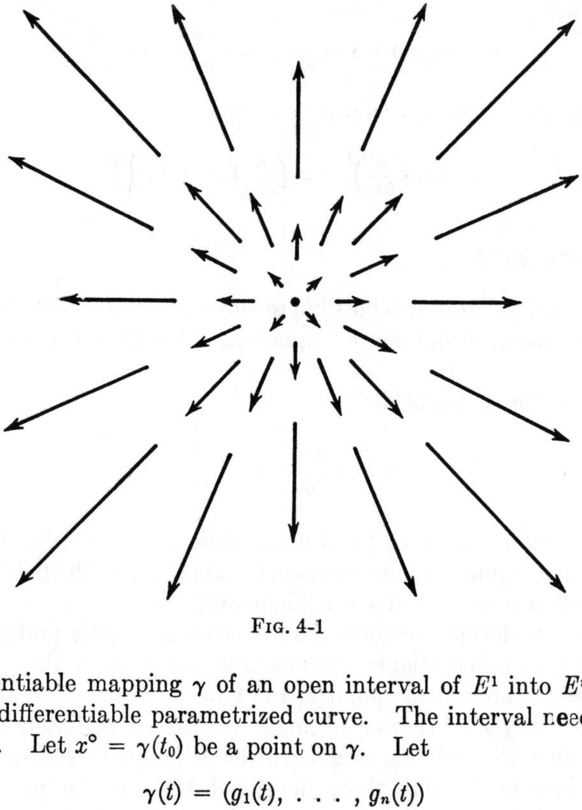

FIG. 4-1

A differentiable mapping γ of an open interval of E^1 into E^n will be called a differentiable parametrized curve. The interval need not be bounded. Let $x^\circ = \gamma(t_0)$ be a point on γ. Let

$$\gamma(t) = (g_1(t), \ldots, g_n(t))$$

Then the tangent vector to γ at x° is

$$g_1'(t_0)\left(\frac{\partial}{\partial x_1}\right)_{x^\circ} + \cdots + g_n'(t_0)\left(\frac{\partial}{\partial x_n}\right)_{x^\circ}$$

Let χ be a differentiable vector field on an open subset U of E^n. A differentiable parametrized curve γ will be called an *integral manifold* for χ if the tangent vector to γ at the point $\gamma(t)$ is $\chi(\gamma(t))$ for each t for which γ is defined.

The collection of all integral manifolds for χ is called the *flow* determined by χ, by analogy with the lines of flow of the velocity field of a moving fluid. Let $\gamma(t) = (x_1(t), \ldots, x_n(t))$ be an integral manifold for the differentiable vector field

$$\chi = a_1 \frac{\partial}{\partial x_1} + \cdots + a_n \frac{\partial}{\partial x_n}$$

on the open set U. Then $a_1, \ldots, a_n$ are differentiable functions of $x_1, \ldots, x_n$ on U. The tangent vector to γ at $\gamma(t_0)$ is

$$x_1'(t_0) \frac{\partial}{\partial x_1} + \cdots + x_n'(t_0) \frac{\partial}{\partial x_n}$$

We therefore find that the functions $x_1(t), \ldots, x_n(t)$ must satisfy the system

$$\begin{aligned} x_1'(t) &= a_1(x_1, \ldots, x_n) \\ &\cdots\cdots\cdots\cdots \\ x_n'(t) &= a_n(x_1, \ldots, x_n) \end{aligned}$$

of differential equations. Conversely, if there is a solution $x_1(t), \ldots, x_n(t)$ of this system of equations, the curve γ given by

$$\gamma(t) = (x_1(t), \ldots, x_n(t))$$

is an integral manifold for the vector field χ.

Returning to our example, the integral manifolds for

$$\chi = x_1 \frac{\partial}{\partial x_1} + x_2 \frac{\partial}{\partial x_2}$$

on E^2 are the solutions of the system

$$x_1'(t) = x_1 \qquad x_2'(t) = x_2$$

These are $x_1 = c_1e^t$, $x_2 = c_2e^t$. Since $x_2 = (c_2/c_1)x_1$, we see that, as we would expect, the only integral manifolds are the straight lines radiating from the origin, plus the trivial case when $\gamma(t) = (0, 0)$ for all t. In this example, the origin is a singularity, for none of the nontrivial integral manifolds pass through it.

The question of the existence and uniqueness of the integral manifolds for an arbitrary differentiable vector field is settled by means of the Picard theorem on the existence and uniqueness of the solutions of systems of first-order differential equations. For the proof of this theorem, the reader is referred to L. R. Ford, "Differential Equations" (2d ed., chaps. 5, 6, McGraw-Hill, New York, 1955) (it is called the method of successive approximations in this book). We shall state the theorem in a weakened form which is, however, sufficient for the study of integral manifolds.

Picard Theorem. *Let* $a_1, \ldots, a_n$ *be differentiable functions on a neighborhood* U *of the point* y° *in* E^n. *Then there are a neighborhood* V *of* y° *contained in* U, *a positive number* ϵ, *and unique functions* $x_1(y, t), \ldots, x_n(y, t)$, *which are differentiable for* $y \in V$ *and* $|t| < \epsilon$, *such that* $y = (x_1(y, 0), \ldots, x_n(y, 0))$, *the point* $(x_1(y, t), \ldots, x_n(y, t))$ *is in* U, *and*

$$\frac{\partial x_1}{\partial t} = a_1(x_1(y, t), \ldots, x_n(y, t))$$
$$\cdots\cdots\cdots\cdots\cdots\cdots$$
$$\frac{\partial x_n}{\partial t} = a_n(x_1(y, t), \ldots, x_n(y, t))$$

when $y \in V$ *and* $|t| < \epsilon$.

Let

$$\chi = a_1 \frac{\partial}{\partial x_1} + \cdots + a_n \frac{\partial}{\partial x_n}$$

be a differentiable vector field on an open subset U of E^n. Let y° be a point in U. The Picard theorem gives us a neighborhood V of y° which is contained in U, a positive number ϵ, and a family

$$\gamma(y, t) = (x_1(y, t), \ldots, x_n(y, t))$$

of integral manifolds for χ. The integral manifold $\gamma(y, t)$ lies in U and passes through the point y in V. Furthermore, the integral manifolds are unique in the sense that, if $\tilde{\gamma}(t)$ is any integral manifold for χ such that $\tilde{\gamma}(0) = y \in V$, then $\tilde{\gamma}(t) = \gamma(y, t)$ for $|t| < \epsilon$. Consequently, if $\tilde{\gamma}(t)$ is any integral manifold for χ such that $\tilde{\gamma}(t_0) = y \in V$, then $\tilde{\gamma}(t + t_0) = \gamma(y, t)$ for $|t| < \epsilon$. Finally we observe that the integral manifolds $\gamma(y, t)$ depend differentiably on the point y in V.

If $\chi(y^\circ) = 0$, the integral manifold through y° is the trivial one $\gamma(t) = y^\circ$ for all t. If $\chi(y^\circ) \neq 0$, then the integral manifold through y° must actually pass through y°. The singularities of the flow associated with χ are therefore precisely the points where χ vanishes.

The problem of flows is capable of generalization in the following way. In place of the given vector field we are given a correspondence X which assigns to each point x a 2-dimensional subspace $X(x)$ (or, more generally, a p-dimensional subspace) of the tangent space $T(x)$. An integral manifold is then a 2-dimensional submanifold S of E^n (or a

p-dimensional submanifold) such that at each point x the tangent space to S at x is $\mathrm{X}(x)$. These problems will be taken up in a later chapter.

EXERCISES

1. Determine the flows associated with the following vector fields in E^2:

a.
$$\mathrm{X} = x_2 \frac{\partial}{\partial x_1} - x_1 \frac{\partial}{\partial x_2}$$

b.
$$\mathrm{X} = x_2 \frac{\partial}{\partial x_1} - x_2{}^3 \frac{\partial}{\partial x_2}$$

2. Determine the flow associated with the vector field

$$\mathrm{X} = x_2 \frac{\partial}{\partial x_1} + x_3 \frac{\partial}{\partial x_2} + x_1 \frac{\partial}{\partial x_3}$$

in E^3.

3. Find the vector field of the flow $\{\gamma_{a,t_0}\}$, where

$$\gamma_{a,t_0}(t) = (a \sin (t - t_0), a \cos (t - t_0), a^2 t)$$

4-5 FORMS ON E^n

Let U be an open subset of E^n. A *form* on U is a correspondence ω that assigns to each point x of U a differential $\omega(x)$ from $T^*(x)$. Since the cotangent space at x is dual to the tangent space at x, it is to be expected that the theory of forms will be dual to the theory of vector fields.

In order to define the concept of differentiability for forms, we introduce the cotangent bundle of E^n. Let $\mathfrak{T}^*(U)$ be the set of all pairs (x, λ), where x is in U and λ is in $T^*(x)$. If (x, λ) is a point of $\mathfrak{T}^*(U)$, then

$$\lambda = \sum_{i=1}^{n} c_i (dx_i)_x$$

The mapping

$$(x, \lambda) \to (x_1, \ldots, x_n, c_1, \ldots, c_n)$$

of $\mathfrak{T}^*(U)$ onto the subset $U \times E^n$ of E^{2n} is one-to-one. We give to $\mathfrak{T}^*(U)$ the topology that makes this mapping a homeomorphism, and then we may regard the mapping as a coordinate mapping. In this way $\mathfrak{T}^*(U)$ is covered with one coordinate neighborhood and becomes a differentiable manifold of dimension $2n$. The manifold $\mathfrak{T}^*(U)$ is called the *cotangent bundle* over U.

If ω is a form on U, we may introduce the mapping $x \to (x, \omega(x))$ of U into $\mathfrak{T}^*(U)$. This mapping will also be denoted by ω. We shall say that the form ω is *differentiable* if it is a differentiable mapping of U into $\mathfrak{T}^*(U)$.

For any form ω on U there are functions $a_1, \ldots, a_n$ on U such that

$$\omega(x) = a_1(x)\, dx_1 + \cdots + a_n(x)\, dx_n$$

It is clear that ω is differentiable on U if and only if the functions $a_1, \ldots, a_n$ are all differentiable on U. In particular, the form ω for which $\omega(x) = (dx_i)_x$ for all x is differentiable. It will be denoted by dx_i. If g is a function which is differentiable on U, the form ω for which $\omega(x) = (dg)_x$ for all x is differentiable and will be denoted by dg.

Let $\mathfrak{W}$ be the set of all differentiable forms on U. Let $\mathfrak{D}$ be the ring of all functions which are differentiable on U. Just as with vector fields, we may make $\mathfrak{W}$ a module over $\mathfrak{D}$ by defining

$$(\omega_1 + \omega_2)(x) = \omega_1(x) + \omega_2(x)$$
$$(f\omega_1)(x) = f(x)\omega_1(x)$$

when $\omega_1, \omega_2 \in \mathfrak{W}$ and $f \in \mathfrak{D}$. Then the forms $dx_1, \ldots, dx_n$ span $\mathfrak{W}$ and are linearly independent over $\mathfrak{D}$.

Let $\delta\colon V \to U$ be a one-to-one differentiable mapping of the open subset V of E^n onto an open subset U of E^n such that δ^{-1} is also differentiable. Then δ induces a linear transformation $\delta^*\colon T^*(x) \to T^*(\delta^{-1}(x))$ for each x in U. Let ω be a form on U. If y is a point of V, we may define

$$\omega^\delta(y) = \delta^*[\omega(\delta(y))]$$

Then ω^δ is a form on V which is said to be induced by δ from ω. Let $(x_1, \ldots, x_n) = \delta(y_1, \ldots, y_n)$, and let

$$\omega(x) = \sum_{j=1}^{n} a_j(x)(dx_j)_x$$

There are functions $b_1, \ldots, b_n$ on V such that

$$\omega^\delta(y) = \sum_{i=1}^{n} b_i(y)(dy_i)_y$$

Since

$$\delta^*[\omega(x)] = \sum_{j=1}^{n} a_j(x)\delta^*[(dx_j)_x]$$

and

$$\delta^*[(dx_j)_x] = \sum_{i=1}^{n} \frac{\partial x_j}{\partial y_i}(dy_i)_{\delta^{-1}(x)}$$

it follows that

$$b_i(y) = \sum_{j=1}^{n} \frac{\partial x_j}{\partial y_i} a_j(\delta(y)) \tag{4-5-1}$$

The collection of n-tuples of functions $\{a_1, \ldots, a_n\}$, one collection for each coordinate neighborhood and all related by (4-5-1), is called a covariant vector in the classical literature of differential geometry. The relation (4-5-1) is called the transformation law for a covariant vector. The relation (4-5-1) also tells us that if ω is differentiable on U then ω^δ is differentiable on V.

When a form ω and a vector field χ are defined on the same open set U in E^n, then $\omega(x)$ is a linear functional on $T(x)$ and, in particular, has a value at $\chi(x)$. Thus

$$f(x) = \langle \chi(x), \omega(x) \rangle$$

is a function on U. Let $\omega = \Sigma_{i=1}^n c_i\, dx_i$ and $\chi = \Sigma_{i=1}^n a_i(\partial/\partial x_i)$. Then $f(x) = \Sigma_{i=1}^n c_i(x) a_i(x)$. If ω and χ are differentiable on U, then f will be differentiable on U. It is also clear that a form ω on U is differentiable on U if and only if for every differentiable vector field χ on U the function $\langle \chi, \omega \rangle$ is differentiable on U.

EXERCISES

1. Let

$$\omega_1 = (1 + x_1^2)\, dx_1 + dx_2$$
$$\omega_2 = (x_1^4 + x_2^4)\, dx_1 - (1 - x_1^2)\, dx_2$$

on E^2. Show that ω_1 and ω_2 are linearly independent differentiable forms on E^2. If

$$g(x_1, x_2) = x_1 e^{x_2}$$

find the functions b_1 and b_2 such that $dg = b_1\omega_1 + b_2\omega_2$.

2. Let δ be the inverse of the diffeomorphism of Exercise 7 in Section 1-3. If ω_1 and ω_2 are the forms of the preceding exercise, find ω_1^δ and ω_2^δ.

3. If g_1 and g_2 are two differentiable functions on an open subset U of E^2, find a necessary and sufficient condition that the forms dg_1 and dg_2 be linearly independent on U.

4. Let

$$\omega_1 = e^{x_2}\, dx_1 \qquad \omega_2 = e^{x_3}\, dx_2 \qquad \omega_3 = e^{x_1}\, dx_3$$

on E^3. Show that $\omega_1, \omega_2, \omega_3$ are linearly independent differentiable forms on E^3.

5. Let δ be the inverse of the diffeomorphism of Exercise 8 in Section 1-3. If $\omega_1, \omega_2, \omega_3$ are the forms of the preceding exercise, find $\omega_1^\delta, \omega_2^\delta, \omega_3^\delta$.

6. Let U be an open subset of E^n. Let $\omega_1, \ldots, \omega_n$ be linearly independent differentiable forms on U. Let $\omega = \Sigma_{i=1}^n a_i\omega_i$. Prove that ω is differentiable on U if and only if the functions $a_1, \ldots, a_n$ are all differentiable on U.

7. Let $\omega_1, \ldots, \omega_n$ be the forms of the preceding exercise. Let δ be a diffeomorphism of E^n. Prove that the forms $\omega_1^\delta, \ldots, \omega_n^\delta$ are linearly independent on $\delta^{-1}(U)$.

4-6 THE TANGENT BUNDLE AND COTANGENT BUNDLE OF A DIFFERENTIABLE MANIFOLD

Let M be a differentiable manifold. We may form the set $\mathfrak{T}(M)$ of all pairs (m, α), where $m \in M$ and $\alpha \in T(m)$. We shall give a differentiable structure to the set $\mathfrak{T}(M)$ which is locally the same as the differentiable structure of the tangent bundle defined in Section 4-2.

If (m°, α°) is a pair from $\mathfrak{T}(M)$, there is a coordinate neighborhood U of m° with coordinates $(x_1, \ldots, x_n)$. Let $\pi: \mathfrak{T}(M) \to M$ be the mapping defined by $\pi(m, \alpha) = m$. Then $\pi^{-1}(U)$ is the set of all pairs (m, α) such that $m \in U$. If $(m, \alpha) \in \pi^{-1}(U)$, then $m = (x_1, \ldots, x_n)$ and

$$\alpha = a_1 \frac{\partial}{\partial x_1} + \cdots + a_n \frac{\partial}{\partial x_n}$$

The correspondence

$$(m, \alpha) \to (x_1, \ldots, x_n, a_1, \ldots, a_n)$$

is a one-to-one correspondence between $\pi^{-1}(U)$ and an open subset of E^{2n}. The reader may easily show that there is a unique topology on $\mathfrak{T}(M)$ such that all such mappings, as U runs through the coordinate neighborhoods of M, are homeomorphisms. We would like to use the sets $\pi^{-1}(U)$ as coordinate neighborhoods of $\mathfrak{T}(M)$ with the homeomorphisms

$$(m, \alpha) \to (x_1, \ldots, x_n, a_1, \ldots, a_n)$$

as the coordinate mappings. We must therefore verify that these coordinate neighborhoods are compatible with one another.

Let U and V be two coordinate neighborhoods of M such that $U \cap V$ is not empty. Let $(x_1, \ldots, x_n)$ and $(y_1, \ldots, y_n)$ be coordinates in U and V, respectively. Let ψ be the coordinate transformation on $U \cap V$, so that $y = \psi(x)$. Then ψ is a differentiable mapping. If $(m, \alpha) \in \pi^{-1}(U \cap V)$, then

$$\alpha = \sum_{i=1}^{n} a_i \frac{\partial}{\partial x_i} = \sum_{j=1}^{n} b_j \frac{\partial}{\partial y_j}$$

where
$$\frac{\partial}{\partial x_i} = \sum_{j=1}^{n} \frac{\partial y_j}{\partial x_i} \frac{\partial}{\partial y_j}$$

Consequently

$$b_j = \sum_{i=1}^{n} \frac{\partial y_j}{\partial x_i} a_i$$

The partial derivatives $\partial y_j/\partial x_i$ are differentiable functions of $x_1, \ldots, x_n$, and so the coordinates $b_1, \ldots, b_n$ are differentiable functions of $x_1, \ldots, x_n, a_1, \ldots, a_n$. The coordinate transformation from $(x_1, \ldots, x_n, a_1, \ldots, a_n)$ to $(y_1, \ldots, y_n, b_1, \ldots, b_n)$ is therefore differentiable. This means that our choice of coordinate neighborhoods gives to $\mathfrak{T}(M)$ the structure of a differentiable manifold of dimension $2n$. The manifold $\mathfrak{T}(M)$ is called the *tangent bundle* of the manifold M.

The reader may easily construct for himself the analogous *cotangent bundle* $\mathfrak{T}^*(M)$ of the manifold M.

Let U be an open subset of a differentiable manifold M. A differentiable mapping $\chi: U \to \mathfrak{T}(U)$ such that $\pi \circ \chi =$ identity is called a differentiable vector field on U. We may also look upon χ as a correspondence which assigns to each point x of U a vector $\chi(x)$ in $T(x)$. The set $\mathfrak{X}$ of all differentiable vector fields on U may in the usual way be made into a module over the ring $\mathfrak{D}$ of all differentiable functions on U. We leave to the reader the analogous definition of a differentiable form on U.

Let χ be a differentiable vector field on an open subset U of a differentiable manifold M. An *integral manifold* for χ is a differentiable mapping $\gamma: J \to U$, where J is an open interval of E^1, such that

$$\gamma_* \left(\frac{\partial}{\partial t}\right) = \chi(\gamma(t))$$

for all points t in J. Here γ_* denotes the linear transformation induced on the tangent space by the differentiable mapping γ. When U is a coordinate neighborhood of M with the coordinates $(x_1, \ldots, x_n)$, we may write

$$\gamma(t) = (x_1(t), \ldots, x_n(t))$$

where $x_1, \ldots, x_n$ are now differentiable functions of t on the interval J. We may also write

$$\chi(x_1, \ldots, x_n) = a_1 \frac{\partial}{\partial x_1} + \cdots + a_n \frac{\partial}{\partial x_n}$$

where $a_1, \ldots, a_n$ are differentiable functions of $x_1, \ldots, x_n$. Then γ is an integral manifold for χ if and only if $x_1(t), \ldots, x_n(t)$ are solutions of the system of differential equations

$$x_1'(t) = a_1(x_1, \ldots, x_n)$$
$$\cdots\cdots\cdots\cdots\cdots$$
$$x_n'(t) = a_n(x_1, \ldots, x_n)$$

The question of the existence and uniqueness of integral manifolds in this case is therefore no different from the same question for E^n which was settled in Section 4-4.

EXERCISES

1. Let M be the 2-torus. Construct two differentiable vector fields χ_1, χ_2 on M which are linearly independent.

2. Construct two differentiable forms on the manifold M of the preceding exercise which are linearly independent.

4-7 THE PRINCIPAL BUNDLE

Let M be a differentiable manifold of dimension n. As a point set, the *principal bundle* $\mathfrak{P}(M)$ is the set of all pairs $(m, \{\alpha_1, \ldots, \alpha_n\})$, where $m \in M$ and $\{\alpha_1, \ldots, \alpha_n\}$ is a basis of the tangent space $T(m)$ at m. We shall give $\mathfrak{P}(M)$ a differentiable structure by a procedure analogous to the construction of $\mathfrak{T}(M)$ in Section 4-6.

Let $\pi: \mathfrak{P}(M) \to M$ be defined by $\pi(m, \{\alpha_1, \ldots, \alpha_n\}) = m$. Let U be a coordinate neighborhood of M, with the coordinates $x_1, \ldots, x_n$. We shall make $\pi^{-1}(U)$ into a coordinate neighborhood by defining a homeomorphism φ of $\pi^{-1}(U)$ into E^{n+n^2}. The mapping φ is given by

$$\varphi(m, \{\alpha_1, \ldots, \alpha_n\}) = (x_1, \ldots, x_n, a_{11}, a_{12}, \ldots, a_{nn})$$

where

$$\alpha_i = a_{i1}\left(\frac{\partial}{\partial x_1}\right)_m + \cdots + a_{in}\left(\frac{\partial}{\partial x_n}\right)_m$$

for $i = 1, \ldots, n$. The verification that the sets $\pi^{-1}(U)$, together with their mappings φ, provide $\mathfrak{P}(M)$ with the structure of a differentiable manifold will be left to the reader.

EXERCISES

1. Carry out the construction of $\mathfrak{P}(M)$ which is sketched in this section.

2. Let M be a differentiable manifold. Let $\mathfrak{P}^*(M)$ consist of all pairs (m, σ^*), where m is a point of M and $\sigma^* \in GL(T^*(m))$. Show how to make $\mathfrak{P}^*(M)$ into a differentiable manifold. Show that there is a differentiable mapping δ of $\mathfrak{P}(M)$ onto $\mathfrak{P}^*(M)$ such that the following conditions are satisfied:

a. δ is one-to-one.

b. δ^{-1} is also differentiable.

c. $\pi^* \circ \delta = \pi$, where π and π^* are the projection mappings given by $\pi(m, \sigma) = m$ and $\pi^*(m, \sigma^*) = m$, respectively.

CHAPTER 5
Submanifolds and Riemann Metrics

5-1 SUBMANIFOLDS

The general notion of a submanifold of a differentiable manifold is suggested by the example of a differentiable variety in E^n. In general terms, a subset M' of a differentiable manifold M is a submanifold if it is a differentiable manifold whose structure is derived from the differentiable structure of M. This means that the differentiable functions on M' must be locally the restrictions of differentiable functions on M.

Theorem 5-1. *Let M and N be differentiable manifolds. Let $\sigma: N \to M$ be a differentiable mapping, and let $n \in N$. If for every differentiable function f on N there are a neighborhood U of n and a differentiable function g on M such that $g \circ \sigma = f$ on U, then σ is regular at n.*

Proof. Let $n' = \sigma(n)$. Let V be a coordinate neighborhood of n and $(x_1, \ldots, x_r)$ the coordinates on V. There are functions $f_1, \ldots, f_r$ on V such that $dx_i = df_i$ at n for $i = 1, \ldots, r$. By the hypotheses of this theorem there is a neighborhood U of n and there are functions $g_1, \ldots, g_r$ on M such that $f_i = g_i \circ \sigma$ on U for $i = 1, \ldots, r$. It follows that if σ^* is the linear transformation of $T^*(n')$ induced by σ then $dx_i = \sigma^*(dg_i)$ at n for $i = 1, \ldots, r$.

Suppose that $\alpha \in T(n)$ and $\sigma_*(\alpha) = 0$, where σ_* is the linear transformation of $T(n)$ induced by σ. Then $\langle \sigma_*(\alpha), dg_j \rangle = 0$ for $j = 1, \ldots, r$. But then $\langle \alpha, \sigma^*(dg_j) \rangle = 0$ or $\langle \alpha, dx_j \rangle = 0$ for $j = 1, \ldots, r$. It follows that α must be 0; hence σ_* is a one-to-one mapping, which was to be proved.

Let M be a differentiable manifold, and let M' be a subset of M. M' is called a *submanifold* of M if there are a differentiable manifold N and a differentiable mapping $\sigma: N \to M$ such that the following conditions are satisfied:

1. σ is one-to-one.
2. $\sigma(N) = M'$.
3. For any differentiable function f on N and any point n in N there are a neighborhood U of n and a differentiable function g on M such that $g \circ \sigma = f$ on U.

If the differentiable structure of N is transferred to M' by the one-to-one mapping σ, the differentiable functions on M' are precisely those functions which are locally the restrictions of differentiable functions on M. On the other hand, in Section 5-2 we shall see that the topology which is transferred to M' from N need not be the topology induced on M' from M.

As one can see from Theorems 2-4 and 2-5, the converse of Theorem 5-1 is also true.

Theorem 5-2. *If M and N are differentiable manifolds and if $\sigma: N \to M$ is a regular differentiable mapping, then for any point n in N and any differentiable function f on N there are a neighborhood U of n and a differentiable function g on M such that $g \circ \sigma = f$ on U.*

Thus a subset M' of M is a submanifold of M if there are a differentiable manifold N and a one-to-one regular differentiable mapping $\sigma: N \to M$ such that $M' = \sigma(N)$. It is clear that, if M' is a subset of a differentiable manifold M which can be made into a submanifold of M, its differentiable structure is uniquely determined by the differentiable structure of M.

Let M' be a submanifold of M. Let $\sigma: M' \to M$ be defined by $\sigma(m) = m$. Then σ is a regular differentiable mapping. If $m \in M'$, there are a neighborhood V of $\sigma(m)$ in M and a differentiable mapping $\tau: V \to M'$ such that $U = \tau(V)$ is a neighborhood of m in M' and $\tau \circ \sigma =$ identity on U. It follows that $\tau(m) = m$ for the points m in V and $U = V \cap M'$. If g is a differentiable function on V, then $f = g \circ \sigma$ is the restriction of g to M'. In terms of the induced linear transformation σ^* on the cotangent space at $\sigma(m)$, we have $df = \sigma^*(dg)$.

On the other hand, let γ be a differentiable parametrized curve in M' through the point m. Then $\sigma \circ \gamma$ is the same differentiable parametrized curve but now regarded as a curve in M. In terms of the induced linear transformations on the tangent spaces, we have $(\sigma \circ \gamma)_* = \sigma_* \circ \gamma_*$. But $\alpha = \gamma_*(\partial/\partial t)$ is the tangent vector to γ at m in M', and $(\sigma \circ \gamma)_*(\partial/\partial_t)$ is the tangent vector to $\sigma \circ \gamma$ at m in M. Consequently we find that *the tangent vector to $\sigma \circ \gamma$ at m is the image under σ_* of the tangent vector to γ at m.* The mapping σ_* is one-to-one. If we adopt the geometric view of a tangent vector as a family of differentiable curves, the tangent vectors to M' at m may be regarded as those tangent vectors to M at m that are represented by curves which lie in M'. In particular, we may regard the tangent space to M' at m as a subspace of the tangent space to M at m. From the algebraic point of view, we are simply identifying the tangent space to M' at m with its image under σ_*.

Now let λ be a differential at m in M. If α is any tangent vector to M' at m, we have $\langle \alpha, \sigma^*(\lambda) \rangle = \langle \sigma_*(\alpha), \lambda \rangle$. In other words, if we identify α and $\sigma_*(\alpha)$, the value of $\sigma^*(\lambda)$ at α is the same as the value of λ at α. Since $\sigma^* \circ \tau^* = (\tau \circ \sigma)^* =$ identity, σ^* maps the cotangent space of M at m onto the cotangent space of M' at m. Consequently we may say that *the differentials of M' arise from the differentials of M by restricting them to the tangent space to M'.*

EXERCISES

1. Prove Theorem 5-2.

2. Prove that the differentiable varieties in E^n are submanifolds of E^n.

3. Prove that if M' is a submanifold of M and if M'' is a submanifold of M', then M'' is a submanifold of M.

4. Prove that every open subset of a differentiable manifold M is a submanifold of M.

5. Let M be a differentiable manifold and M' a submanifold of M. Let N be a differentiable manifold and $\mu: N \to M'$ be a continuous mapping. Prove that μ is differentiable as a mapping into M' if and only if μ is differentiable as a mapping into M.

6. Under the hypotheses of the previous exercise, let $\tau: M' \to N$ be a mapping. Prove that τ is differentiable if and only if τ is locally the restriction to M' of differentiable mappings from M into N.

7. Let M' be a submanifold of the differentiable manifold M. Let $\mathfrak{T}(M')$ and $\mathfrak{T}(M)$ be the tangent bundles of M' and M, respectively. Show how $\mathfrak{T}(M')$ may be regarded as a submanifold of $\mathfrak{T}(M)$.

5-2 AN EXAMPLE

In this section we shall consider an example which shows that it is possible for a differentiable manifold M to have a submanifold *of lower dimension* which is everywhere dense in M.

Let T^2 be the 2-torus described in Section 2-4. To simplify the computations, choose the basis $\{v^1 = (1, 0), v^2 = (0, 1)\}$ for V^2. Then the points (x_1, x_2) and (y_1, y_2) represent the same point of T^2 if and only if $x_1 - y_1$ and $x_2 - y_2$ are integers. Let π be the natural homomorphism of the group V^2 onto the group T^2. Let c be a number, and let L be the differentiable variety in E^2 defined by the equation $x_2 - cx_1 = 0$. Let $S = \pi(L)$. We shall see that S is a submanifold of T^2.

Suppose that $\pi: V^2 \to T^2$ is not one-to-one when restricted to L. Then there are distinct numbers t_1 and t_2 such that $\pi(t_1, ct_1) = \pi(t_2, ct_2)$. Thus there are integers l and m such that $t_1 - t_2 = m$ and $ct_1 - ct_2 = l$. Since m is not zero, $c = l/m$ is a rational number. We may assume that the fraction l/m is in lowest terms. Let S^1 be the identification space model of the 1-sphere, and let $\theta: E^1 \to S^1$ be the identification mapping. In other words, $\theta(t_1) = \theta(t_2)$ if and only if $t_1 - t_2$ is an integer. Then the mapping $\theta(t) \to \pi(mt, lt)$ of S^1 into T^2 is well-defined and is actually a one-to-one mapping of S^1 onto S. Furthermore, this mapping is differentiable and regular, and so S is a submanifold of T^2 with the differentiable structure of S^1.

Henceforth we shall assume that c is an irrational number, so that $\pi: V^2 \to T^2$ is one-to-one on L. Thus the mapping $t \to \pi(t, ct)$ is a one-to-one mapping of E^1 onto S which is differentiable and regular. This shows that S is a submanifold of T^2 with the differentiable structure of E^1. We shall prove that S is everywhere dense in T^2.

The set $\pi^{-1}(S)$ consists of all points $(t + l, ct + m)$, where l and m are integers and t is any real number. It therefore consists of the union of all straight lines in E^2 whose equations have the form

$$x_2 - m - c(x_1 - l) = 0$$

where m and l are integers. If we prove that $\pi^{-1}(S)$ is everywhere dense in E^2, it will follow that S is everywhere dense in T^2. Since the lines of $\pi^{-1}(S)$ are all parallel to one another, it suffices to show that their intersections with the line $x_1 = 0$ are everywhere dense in this line. These intersection points have the form $(0, m - cl)$, where m and l are integers. We must therefore prove that, if y is any number and ϵ is any positive number, there are integers m and l such that $|m - cl - y| < \epsilon$.

We shall first show that it suffices to establish this statement when $y = 0$. For suppose that we are able to find integers l_1 and m_1 such that $|m_1 - cl_1| < \epsilon$. Then $c_1 = m_1 - cl_1$ is an irrational number, and so c_1 is not zero. Let k be the largest integer which does not exceed y/c_1. Then let $m = km_1$ and $l = kl_1$. It follows that

$$|m - cl - y| = |kc_1 - y| = |c_1| \left| k - \frac{y}{c_1} \right| < |c_1| < \epsilon$$

which was to be proved.

The case when $y = 0$ follows easily from the following theorem concerning the approximation of irrational numbers by rational numbers.

Theorem 5-3. *Let c be an irrational number. Then there are sequences of integers $m_1, m_2, \ldots$ and $n_1, n_2, \ldots$ such that $n_i \geqq i - 2$ and*

$$\left| c - \frac{m_i}{n_i} \right| \leqq \frac{1}{n_i^2}$$

for $i = 2, 3, \ldots$.

Proof. Define the sequence $c_0, c_1, c_2, \ldots$ of irrational numbers and the sequence $l_1, l_2, \ldots$ of integers as follows: $c_0 = c$, l_i is the largest integer which does not exceed c_{i-1}, and

$$c_i = \frac{1}{c_{i-1} - l_i}$$

Since $0 < c_{i-1} - l_i < 1$, it follows that $c_i > 1$ and hence that $l_{i+1} \geqq 1$ for $i = 1, 2, \ldots$. Then define two sequences of integers $m_1, m_2, \ldots$ and $n_1, n_2, \ldots$ as follows: $m_1 = 1$, $m_2 = l_1$, $n_1 = 0$, $n_2 = 1$, and

$$\begin{aligned} m_{i+1} &= l_i m_i + m_{i-1} \\ n_{i+1} &= l_i n_i + n_{i-1} \end{aligned}$$

for $i = 2, 3, \ldots$. In particular,

$$\begin{aligned} n_3 &= l_2 n_2 + n_1 = l_2 \geqq 1 \\ n_4 &= l_3 n_3 + n_2 = l_3 l_2 + 1 \geqq 2 \end{aligned}$$

Once we have proved that $n_j \geqq j - 2$ for $j < i$ with $i \geqq 5$, it follows that

$$n_i = l_{i-1} n_{i-1} + n_{i-2} \geqq i - 3 + i - 4 \geqq i - 2$$

since $i - 3 \geqq 2$. Hence this relation is established by induction for all i.

We shall prove by induction that

$$m_i n_{i-1} - n_i m_{i-1} = (-1)^{i-1}$$

for $i = 2, 3, \ldots$. The case $i = 2$ is easily settled by substitution. Furthermore,

$$\begin{aligned} m_i n_{i-1} - n_i m_{i-1} &= (l_{i-1} m_{i-1} + m_{i-2}) n_{i-1} - (l_{i-1} n_{i-1} + n_{i-2}) m_{i-1} \\ &= -(m_{i-1} n_{i-2} - n_{i-1} m_{i-2}) \end{aligned}$$

from which the general case follows. The relation may be rewritten as

$$\frac{m_{i-1}}{n_{i-1}} - \frac{m_i}{n_i} = \frac{(-1)^i}{n_{i-1} n_i}$$

Finally we shall prove by induction that

$$c - \frac{m_i}{n_i} = \frac{(-1)^i}{n_i (n_i c_{i-1} + n_{i-1})}$$

for $i = 2, 3, \ldots$. The case $i = 2$ may be verified by substitution.

If we assume that

$$c - \frac{m_{i-1}}{n_{i-1}} = \frac{(-1)^{i-1}}{n_{i-1}(n_{i-1}c_{i-2} + n_{i-2})}$$

then
$$\begin{aligned} c - \frac{m_i}{n_i} &= c - \frac{m_{i-1}}{n_{i-1}} + \frac{m_{i-1}}{n_{i-1}} - \frac{m_i}{n_i} \\ &= \frac{(-1)^i}{n_{i-1}n_i} \frac{n_{i-1}c_{i-2} - (n_i - n_{i-2})}{n_{i-1}c_{i-2} + n_{i-2}} \\ &= \frac{(-1)^i}{n_i} \frac{c_{i-2} - l_{i-1}}{n_{i-1}c_{i-2} + n_{i-2}} \\ &= \frac{(-1)^i}{n_i} \frac{1}{c_{i-1}(n_{i-1}c_{i-2} + n_{i-2})} \end{aligned}$$

Since $c_{i-1}c_{i-2} = c_{i-1}l_{i-1} + 1$, we obtain

$$\begin{aligned} c - \frac{m_i}{n_i} &= \frac{(-1)^i}{n_i} \frac{1}{(l_{i-1}n_{i-1} + n_{i-2})c_{i-1} + n_{i-1}} \\ &= \frac{(-1)^i}{n_i(n_ic_{i-1} + n_{i-1})} \end{aligned}$$

From the inequalities $c_{i-1} \geqq 1$ and $n_{i-1} \geqq 0$ for $i = 2, 3, \ldots$ it follows that

$$n_i(n_ic_{i-1} + n_{i-1}) \geqq n_i^2$$

Hence

$$\left| c - \frac{m_i}{n_i} \right| \leqq \frac{1}{n_i^2}$$

EXERCISE

The 3-torus T^3 is obtained from V^3 in the same way that T^2 is obtained from E^2, namely, by identifying two vectors (x_1, x_2, x_3) and (y_1, y_2, y_3) for which $x_1 - y_1$, $x_2 - y_2$, and $x_3 - y_3$ are integers. Let π: $V^3 \to T^3$ be the identification mapping. Let L be the differentiable variety in $E^3 = V^3$ defined by the equations $x_2 - c_1x_1 = x_3 - c_2x_1 = 0$. Prove that $S = \pi(L)$ is a submanifold of T^3. Prove that S is dense in T^3 if and only if c_1, c_2, and c_1/c_2 are irrational.

5-3 PRODUCTS OF MANIFOLDS

Let M and M' be differentiable manifolds of dimensions n and n', respectively. The topological product $M \times M'$ is a Hausdorff space with a denumerable basis. We shall see that $M \times M'$ may be given the structure of a differentiable manifold in a natural way.

Let $\{U_\alpha\}$ be the family of all coordinate neighborhoods of M and $\{U'_{\alpha'}\}$ the family of all coordinate neighborhoods of M'. Let φ_α: $U_\alpha \to E^n$ and $\varphi'_{\alpha'}$: $U'_{\alpha'} \to E^{n'}$ be the corresponding coordinate mappings. The open subsets $\{U_\alpha \times U'_{\alpha'}\}$ cover $M \times M'$, and the mapping $\varphi_\alpha \times \varphi'_{\alpha'}: U_\alpha \times U'_{\alpha'} \to E^n \times E^{n'} = E^{n+n'}$ defined by

$$(\varphi_\alpha \times \varphi'_{\alpha'})(m, m') = (\varphi_\alpha(m), \varphi'_{\alpha'}(m'))$$

is a homeomorphism of $U_\alpha \times U'_{\alpha'}$ onto an open subset of $E^{n+n'}$. Furthermore, since

$$(\varphi_\beta \times \varphi'_{\beta'})^{-1} \circ (\varphi_\alpha \times \varphi'_{\alpha'}) = \varphi_\beta^{-1} \circ \varphi_\alpha \times \varphi'^{-1}_{\beta'} \circ \varphi'_{\alpha'}$$

it follows that this mapping is differentiable on the set in $E^{n+n'}$ on which it is defined. Consequently the sets $U_\alpha \times U'_{\alpha'}$ become coordinate neighborhoods of $M \times M'$ with the corresponding coordinate mappings $\varphi_\alpha \times \varphi'_{\alpha'}$.

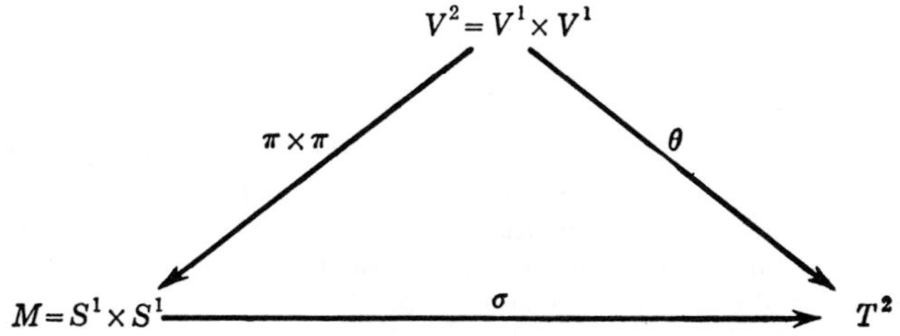

FIG. 5-1

Various properties which are true of topological products and continuous mappings also hold for products of differentiable manifolds and differentiable mappings. These properties are listed in the exercises and their proofs are left to the reader.

As an example, we shall show that if S^1 is the 1-sphere then $S^1 \times S^1$ is a differentiable manifold with the same structure as the 2-torus T^2. This gives us a third model for the 2-torus.

S^1 can be regarded as the differentiable variety in E^2 defined by $x_1^2 + x_2^2 = 1$. However, we shall use the identification space model in which S^1 is obtained by identifying any two points t and t' of V^1 such that $t - t'$ is an integer. The details of this construction are similar to those for the 2-torus and will be left to the reader. (Of course, S^1 is also the 1-torus T^1.) Let $\pi: V^1 \to S^1$ be the identification mapping. Then π is differentiable.

Let M denote the manifold $S^1 \times S^1$, and consider the differentiable mapping $\pi \times \pi: V^1 \times V^1 = V^2 \to M$ defined by

$$(\pi \times \pi)(t, t') = (\pi(t), \pi(t'))$$

Let $\theta: V^2 \to T^2$ be the identification mapping. It is clear that there is a one-to-one mapping σ of M onto T^2 (see Figure 5-1) such that

$\sigma\circ(\pi \times \pi) = \theta$. We wish to show that σ and σ^{-1} are differentiable mappings. Then M and T^2 will have the same differentiable structure.

Let $m = (s_1, s_2) \in M$. The linear transformation π_* which is induced by $\pi\colon V^1 \to S^1$ is onto. By Theorem 2-6, there are neighborhoods U_1 and U_2 of s_1 and s_2, respectively, and differentiable mappings $\tau_1\colon U_1 \to V^1$ and $\tau_2\colon U_2 \to V^1$ such that $\pi\circ\tau_1 =$ identity on U_1 and $\pi\circ\tau_2 =$ identity on U_2. Then

$$\theta\circ(\tau_1 \times \tau_2) = \sigma\circ(\pi \times \pi)\circ(\tau_1 \times \tau_2) = \sigma$$

on $U_1 \times U_2$. Hence σ is differentiable at m.

Since $\theta\colon V^2 \to T^2$ induces a linear transformation θ_* which is onto, we may prove in the same fashion that σ^{-1} is differentiable.

EXERCISES

1. Let M, M', and M'' be differentiable manifolds. Let $\mu\colon (M \times M') \times M'' \to M \times (M' \times M'')$ be the natural homeomorphism of these topological products. Prove that μ and μ^{-1} are differentiable mappings. This shows that these two products of manifolds have the same differentiable structure. It is customary to identify them and denote either one by $M \times M' \times M''$.

2. Let M and M' be differentiable manifolds. Let $\pi\colon M \times M' \to M$ be the natural projection. Prove that π is a differentiable mapping.

3. Let U and U' be open subsets of differentiable manifolds. Let M and M' be differentiable manifolds. Let $\delta\colon U \to M$ and $\delta'\colon U' \to M'$ be differentiable mappings. Let $\delta \times \delta'\colon U \times U' \to M \times M'$ be defined in the natural way, and prove that $\delta \times \delta'$ is differentiable.

4. In Exercise 3, prove that $\delta \times \delta'$ is regular at the point (u, u') of $U \times U'$ if and only if δ is regular at u and δ' is regular at u'.

5. Let M and M' be differentiable manifolds. Let $m \in M$ and $m' \in M'$, and let $T(m)$, $T(m')$, and $T(m, m')$ denote the tangent spaces at m, m', and $(m, m') \in M \times M'$, respectively. Prove that $T(m, m')$ is isomorphic in a natural way to $T(m) \oplus T(m)'$, that is, every element of $T(m, m')$ may be written in one and only one way as $\alpha + \alpha'$ with $\alpha \in T(m)$ and $\alpha' \in T(m')$.

6. Let M and M' be differentiable manifolds. Let $\mathfrak{T}(M)$, $\mathfrak{T}(M')$, and $\mathfrak{T}(M \times M')$ be the tangent bundles of M, M', and $M \times M'$, respectively. Prove that $\mathfrak{T}(M \times M')$ has the differentiable structure of $\mathfrak{T}(M) \times \mathfrak{T}(M')$.

7. Prove statements similar to Exercises 5 and 6 for the cotangent space and the cotangent bundle.

5-4 RIEMANN METRICS

Let x be a point in E^2. Let $T(x)$ be the tangent space at x. We may introduce an inner product in $T(x)$ as follows: If

$$\alpha = a_1\left(\frac{\partial}{\partial x_1}\right)_x + a_2\left(\frac{\partial}{\partial x_2}\right)_x$$

and
$$\beta = b_1 \left(\frac{\partial}{\partial x_1}\right)_x + b_2 \left(\frac{\partial}{\partial x_2}\right)_x$$

then
$$\langle \alpha, \beta \rangle = a_1 b_1 + a_2 b_2$$

The inner product assigns a length $\sqrt{\langle \alpha, \alpha \rangle}$ to each vector α in $T(x)$. We shall introduce this inner product into the tangent space at each point of E^2.

Let γ be a differentiable parametrized curve in E^2. Let

$$\gamma(t) = (f_1(t), f_2(t))$$

for $a \leqq t \leqq b$. The ordinary arc length s along γ is given by

$$s(t) = \int_a^t \sqrt{f_1'(u)^2 + f_2'(u)^2}\, du = \int_a^t \sqrt{\langle \alpha(u), \alpha(u) \rangle}\, du$$

where $\alpha(t)$ is the tangent vector to γ at $\gamma(t)$. If γ is not stationary at any point $a < t < b$, that is, if $\alpha(t) \neq 0$ for $a < t < b$, then

$$\langle \alpha(t), \alpha(t) \rangle \neq 0$$

for $a < t < b$. Under these circumstances, $s(t)$ is a differentiable monotone increasing function and has a differentiable inverse $t(s)$ defined for $0 \leqq s \leqq s(b)$. We may therefore easily reparametrize γ by setting $\tilde{\gamma}(s) = \gamma(t(s))$ for $0 \leqq s \leqq s(b)$. That is to say, $\tilde{\gamma}$ and γ have the same points, but these points are parametrized differently. It follows that the tangent vector to $\tilde{\gamma}$ at each point $\tilde{\gamma}(s)$ with $0 < s < s(b)$ has length 1. Actually, the arc length $s(t)$ along γ may be characterized as that reparametrization of γ for which all the tangent vectors have length 1.

We propose to generalize this situation by permitting other inner products to be introduced into the tangent spaces besides the one which has been discussed. Furthermore, we shall allow the inner product $\langle \alpha, \beta \rangle_m$ to vary in a suitably differentiable manner from one point m of the manifold to another. $\langle \alpha, \beta \rangle_m$ is then called a Riemann metric. Under these circumstances, we shall find that we may still reparametrize a differentiable curve γ in such a way that the tangent vectors all have length 1. This reparametrization will be called the arc length along γ with respect to the Riemann metric $\langle \alpha, \beta \rangle_m$.

Let M be a differentiable manifold. Let $\langle \alpha, \beta \rangle_m$ be a real number which is defined for each point m in an open subset U of M and for all vectors α and β in the tangent space $T(m)$. For each point m in U, $\langle \alpha, \beta \rangle_m$ is assumed to be an inner product on $T(m)$. Let $\xi_1, \ldots, \xi_r$ be vector fields on U. $\xi_1, \ldots, \xi_r$ will be said to be *orthonormal on U* if for each point m in U the set $\{\xi_1(m), \ldots, \xi_r(m)\}$ is orthonormal, that is,

$$\langle \xi_i(m), \xi_j(m) \rangle_m = \begin{cases} 0 & \text{if } i \neq j \\ 1 & \text{if } i = j \end{cases}$$

Theorem 5-4. *Let M be a differentiable manifold with an inner product $\langle \alpha, \beta \rangle_m$ in $T(m)$ at each point m. Let $\chi_1, \ldots, \chi_r$ be linearly independent differentiable vector fields on an open subset U of M. If the functions*

$$g_{ij}(m) = \langle \chi_i(m), \chi_j(m) \rangle_m$$

are differentiable on U for $i, j = 1, \ldots, r$, then there is an orthonormal set $\{\xi_1, \ldots, \xi_r\}$ of differentiable vector fields on U which spans the same module over the ring $\mathfrak{D}$ of differentiable functions on U as the set $\{\chi_1, \ldots, \chi_r\}$.

Proof. We shall apply the orthonormalization procedure described in Theorem 1-4 to the vector fields $\chi_1, \ldots, \chi_r$.

Since $\chi_1(m) \neq 0$ for all $m \in U$, it follows that

$$g_{11}(m) = \langle \chi_1(m), \chi_1(m) \rangle_m \neq 0$$

on U. Hence we may define

$$\xi_1(m) = \frac{1}{\sqrt{g_{11}(m)}} \chi_1(m)$$

on U. Then ξ_1 is a differentiable vector field on U such that

$$\langle \xi_1(m), \xi_1(m) \rangle_m = 1$$

for $m \in U$. Furthermore, ξ_1 spans the same module over $\mathfrak{D}$ as χ_1 and

$$\langle \chi_i(m), \xi_1(m) \rangle_m = \frac{g_{i1}(m)}{\sqrt{g_{11}(m)}}$$

is a differentiable function on U for $i = 1, \ldots, r$.

We have constructed the first vector field ξ_1 of our orthonormal set, and we have proved not only that it is differentiable and spans the same module over $\mathfrak{D}$ as χ_1 but that it satisfies the condition that the functions

$$h_{i1}(m) = \langle \chi_i(m), \xi_1(m) \rangle_m$$

are differentiable on U for $i = 1, \ldots, r$.

We shall now assume that we have constructed differentiable vector fields $\xi_1, \ldots, \xi_s$ on U which are orthonormal, which span the same module over $\mathfrak{D}$ as $\chi_1, \ldots, \chi_s$, and which satisfy the additional condition that for $i = 1, \ldots, r$ and $j = 1, \ldots, s$ the functions

$$h_{ij}(m) = \langle \chi_i(m), \xi_j(m) \rangle_m$$

are differentiable on U. Then we shall show how to construct the vector field ξ_{s+1} on U so that the set $\{\xi_1, \ldots, \xi_{s+1}\}$ satisfies the same conditions. The theorem is then proved by taking s successively equal to $1, \ldots, r-1$.

Consider the vector field ζ defined by

$$\zeta(m) = \chi_{s+1}(m) - \sum_{k=1}^{s} \langle \chi_{s+1}(m), \xi_k(m)\rangle_m \xi_k(m)$$

on U. By our assumptions, ζ is a differentiable vector field on U. Since for each $m \in U$ the vector

$$\sum_{k=1}^{s} \langle \chi_{s+1}(m), \xi_k(m)\rangle_m \xi_k(m)$$

is in the vector space spanned by $\chi_1(m), \ldots, \chi_s(m)$ and $\chi_{s+1}(m)$ is not in this vector space, it follows that $\zeta(m) \neq 0$. Thus we may define

$$\xi_{s+1}(m) = \frac{1}{\sqrt{\langle \zeta(m), \zeta(m)\rangle_m}} \zeta(m)$$

on U.

Since

$$\langle \zeta(m), \zeta(m)\rangle_m = g_{s+1,\,s+1}(m) - \sum_{k=1}^{s} h_{s+1,\,k}(m)^2$$

it follows that $\langle \zeta(m), \zeta(m)\rangle_m$ is differentiable on U. Consequently ξ_{s+1} is a differentiable vector field on U. It is easy to verify that

$$\langle \xi_{s+1}(m), \xi_{s+1}(m)\rangle_m = 1 \qquad \langle \xi_{s+1}(m), \xi_j(m)\rangle_m = 0$$

for $j = 1, \ldots, s$ and that $\xi_1, \ldots, \xi_{s+1}$ span the same module over $\mathfrak{D}$ as $\chi_1, \ldots, \chi_{s+1}$. Finally we observe that

$$h_{i,\,s+1}(m) = \frac{\langle \chi_i(m), \zeta(m)\rangle_m}{\sqrt{\langle \zeta(m), \zeta(m)\rangle_m}} = \frac{g_{i,\,s+1}(m) - \sum_{k=1}^{s} h_{s+1,\,k}(m) h_{ik}(m)}{\sqrt{\langle \zeta(m), \zeta(m)\rangle_m}}$$

and so $h_{i,\,s+1}$ is differentiable on U for $i = 1, \ldots, r$.

We shall say that $\langle \alpha, \beta\rangle_m$ is *differentiable* on U if there is an orthonormal set of n differentiable vector fields on U, where n is the dimension of the manifold M. According to Theorem 5-4, if we can find n linearly independent differentiable vector fields $\chi_1, \ldots, \chi_n$ on U

such that the functions

$$g_{ij}(m) = \langle \chi_i(m), \chi_j(m) \rangle_m$$

are differentiable on U, then $\langle \alpha, \beta \rangle_m$ is differentiable on U. The next theorem shows that the choice of the linearly independent differentiable vector fields $\chi_1, \ldots, \chi_n$ is immaterial.

Theorem 5-5. *If $\langle \alpha, \beta \rangle_m$ is differentiable on an open subset U of a differentiable manifold M and if $\chi_1, \ldots, \chi_r$ are differentiable vector fields on U, then the functions*

$$g_{ij}(m) = \langle \chi_i(m), \chi_j(m) \rangle_m$$

are differentiable on U for $i, j = 1, \ldots, r$.

Proof. There is an orthonormal set $\{\xi_1, \ldots, \xi_n\}$ of differentiable vector fields on U. We may assume that U is a coordinate neighborhood of M with coordinates $(x_1, \ldots, x_m)$. Let

$$\xi_i = \sum_{j=1}^{n} a_{ij} \frac{\partial}{\partial \chi_j} \qquad \chi_i = \sum_{j=1}^{n} b_{ij} \frac{\partial}{\partial \chi_j}$$

Then the functions a_{ij} and b_{ij} are differentiable on U. Since $\xi_1, \ldots, \xi_n$ are linearly independent on U, we may write

$$\chi_i = \sum_{j=1}^{n} c_{ij}\xi_j$$

and it follows that $(b_{ij}) = (c_{ij})(a_{ij})$, in matrix notation. Since (a_{ij}) is nonsingular at all points of U, we have $(c_{ij}) = (b_{ij})(a_{ij})^{-1}$, and so the functions c_{ij} are differentiable on U. Since

$$g_{ij}(m) = \sum_{k=1}^{n} c_{ik}(m)c_{jk}(m)$$

the functions g_{ij} are differentiable on U.

We now see that, in order to test the inner products $\langle \alpha, \beta \rangle_m$ for differentiability on U, it suffices to choose any set of n linearly independent differentiable vector fields on U and to subject these to the orthonormalization procedure. $\langle \alpha, \beta \rangle_m$ is then differentiable if and only if the resulting orthonormal vector fields are differentiable on U. When this is the case we shall say that $\langle \alpha, \beta \rangle_m$ is a *Riemann metric* on U.

Let M' be a submanifold of the differentiable manifold M. Let $\langle \alpha, \beta \rangle_m$ be a Riemann metric on M. Let m be a point in M'. We have seen how we may regard the tangent space $T'(m)$ to M' at m as a subspace of the tangent space $T(m)$ to M at m. Consequently we may restrict $\langle \alpha, \beta \rangle_m$ to $T'(m)$ and in this way obtain a Riemann metric on M'.

Let U be a coordinate neighborhood of M with the coordinates $(x_1, \ldots, x_n)$. Since $\{dx_1, \ldots, dx_n\}$ is a basis of the cotangent space at m, we have

$$\langle \alpha, \beta \rangle_m = \sum_{i=1}^{n} \sum_{j=1}^{n} g_{ij}(m) \langle \alpha, dx_i \rangle \langle \beta, dx_j \rangle$$

(see Section 1-4). Now let V be a coordinate neighborhood of M' which is contained in U. Let the coordinates of V be $(y_1, \ldots, y_r)$. Then the coordinates $x_1, \ldots, x_n$ are differentiable functions of $y_1, \ldots, y_r$, for we know that the mapping $\sigma\colon V \to U$ given by

$$(x_1, \ldots, x_n) = \sigma(y_1, \ldots, y_r)$$

is a regular differentiable mapping. Let m be a point of V. If $\langle \alpha, \beta \rangle'_m$ is the restriction of $\langle \alpha, \beta \rangle_m$ to M', then

$$\langle \alpha, \beta \rangle'_m = \sum_{i=1}^{r} \sum_{j=1}^{r} h_{ij}(m) \langle \alpha, dy_i \rangle \langle \beta, dy_j \rangle$$

The relationship between the functions h_{ij} and the functions g_{ij} is

$$h_{kl}(m) = \sum_{i=1}^{n} \sum_{j=1}^{n} g_{ij}(m) \left(\frac{\partial x_i}{\partial y_k} \frac{\partial x_j}{\partial y_l} \right) \bigg|_m \tag{5-4-1}$$

Let $\langle \alpha, \beta \rangle_m$ be a Riemann metric on an open subset U of a differentiable manifold M. Let γ be a differentiable parametrized curve in U which is never stationary (does not have any zero tangent vectors). We shall show that γ may be reparametrized so that the tangent vectors of the new curve all have length 1 with respect to $\langle \alpha, \beta \rangle_m$.

Let $\gamma(t)$ be defined for $a \leqq t \leqq b$. Let

$$s(t) = \int_a^t \sqrt{\langle \alpha(u), \alpha(u) \rangle_{\gamma(u)}}\, du$$

where $\alpha(t)$ is the tangent vector to γ at $\gamma(t)$. Then $\langle \alpha(t), \alpha(t) \rangle > 0$ for $a < t < b$, and it follows that $s(t)$ is a monotone increasing differentiable function for $a \leqq t \leqq b$. It therefore has an inverse $t(s)$ which is also differentiable and monotone increasing. Let $\tilde{\gamma}(s) = \gamma(t(s))$ for $0 \leqq s \leqq s(b)$. Since the tangent vector to $\tilde{\gamma}$ is

$$t'(s)\alpha(t(s)) = \frac{1}{s'(t)} \alpha(t(s)) = \frac{1}{\sqrt{\langle \alpha(t), \alpha(t) \rangle_{\gamma(t)}}} \alpha(t)$$

it has length 1. The function $s(t)$ is called the arc length along γ with respect to the Riemann metric $\langle \alpha, \beta \rangle_m$.

EXERCISES

1. Let

$$\langle \alpha, \beta \rangle_m = \sum_{i=1}^{n} \sum_{j=1}^{n} g_{ij}(m) \langle \alpha, dx_i \rangle \langle \beta, dx_j \rangle$$

be an assignment of an inner product to the tangent space at each point of a coordinate neighborhood U, with coordinates $(x_1, \ldots, x_n)$. Prove that $\langle \alpha, \beta \rangle_m$ is a Riemann metric on U if and only if all the functions g_{ij} are differentiable on U.

2. Let $\langle \alpha, \beta \rangle_m$ be given as in the preceding exercise. Consider the mapping ρ: $U \times E^{2n} \to E^1$ given by

$$\rho(x_1, \ldots, x_n; a_1, \ldots, a_n; b_1, \ldots, b_n) = \langle \alpha, \beta \rangle_m$$

where

$$m = (x_1, \ldots, x_n)$$

$$\alpha = a_1 \left(\frac{\partial}{\partial x_1}\right)_m + \cdots + a_n \left(\frac{\partial}{\partial x_n}\right)_m$$

$$\beta = b_1 \left(\frac{\partial}{\partial x_1}\right)_m + \cdots + b_n \left(\frac{\partial}{\partial x_n}\right)_m$$

Prove that $\langle \alpha, \beta \rangle_m$ is differentiable on U if and only if the mapping ρ is differentiable.

3. Let

$$g_{11}(x_1, x_2) = x_2^4 + 3x_2^2 + 2x_1x_2 + x_1^2 + 1$$
$$g_{12}(x_1, x_2) = g_{21}(x_1, x_2) = x_2 + x_1x_2^2 + 2x_1$$
$$g_{22}(x_1, x_2) = 2x_1^2 + 1$$

on E^2. Let

$$\langle \alpha, \beta \rangle_x = \sum_{i=1}^{2} \sum_{j=1}^{2} g_{ij}(x) \langle \alpha, dx_i \rangle \langle \beta, dx_j \rangle$$

Show that $\langle \alpha, \beta \rangle_x$ is a Riemann metric on E^2.

4. Orthonormalize the vector fields $\partial/\partial x_1$, $\partial/\partial x_2$ with respect to the Riemann metric of Exercise 3.

5. Let

$$g_{11}(x_1, x_2, x_3) = 2$$
$$g_{12}(x_1, x_2, x_3) = g_{21}(x_1, x_2, x_3) = x_1$$
$$g_{22}(x_1, x_2, x_3) = x_1^2 + 1$$
$$g_{23}(x_1, x_2, x_3) = g_{32}(x_1, x_2, x_3) = x_1$$
$$g_{13}(x_1, x_2, x_3) = g_{31}(x_1, x_2, x_3) = x_2$$
$$g_{33}(x_1, x_2, x_3) = x_1^2 + x_2^2 + 1$$

on E^3. Let

$$\langle \alpha, \beta \rangle_x = \sum_{i=1}^{3} \sum_{j=1}^{3} g_{ij}(x) \langle \alpha, dx_i \rangle \langle \beta, dx_j \rangle$$

Show that $\langle \alpha, \beta \rangle_x$ is a Riemann metric on E^3.

6. Orthonormalize the vector fields $\partial/\partial x_1$, $\partial/\partial x_2$, $\partial/\partial x_3$ with respect to the Riemann metric of Exercise 5.

7. Let M' be a submanifold of the differentiable manifold M. Let U be a coordinate neighborhood of M with the coordinates $(x_1, \ldots, x_n)$. Let V be a coordinate neighborhood of M' with the coordinates $(y_1, \ldots, y_r)$. Let λ_i be the restriction of dx_i to the tangent space to M'. Prove that

$$\lambda_i = \sum_{j=1}^{r} \frac{\partial x_i}{\partial y_j} dy_j$$

and hence derive the relations (5-4-1).

8. Let $\langle \alpha, \beta \rangle_m$ be a Riemann metric on a differentiable manifold M. Let γ be a differentiable parametrized curve in M defined for $a \leqq t \leqq b$. Prove that the following conditions characterize the function $s(t)$ completely:

a. $s(t)$ is a monotone increasing differentiable function for $a < t < b$.

b. $s(a) = 0$.

c. The reparametrization $\tilde{\gamma}(s) = \gamma(t(s))$ produces a differentiable parametrized curve whose tangent vectors with respect to $\langle \alpha, \beta \rangle_m$ all have length 1.

9. Let

$$\langle \alpha, \beta \rangle_m = \sum_{i=1}^{n} \sum_{j=1}^{n} g_{ij}(m) \langle \alpha, dx_i \rangle \langle \beta, dy_j \rangle$$

be a Riemann metric on a coordinate neighborhood U, with coordinates $(x_1, \ldots, x_n)$. Let γ be a differentiable parametrized curve in U without stationary points defined by

$$\gamma(t) = (f_1(t), \ldots, f_n(t))$$

for $a \leqq t \leqq b$. Compute the arc length along γ in terms of the functions g_{ij} and f_i.

10. Let

$$g_{11}(x_1, x_2) = 2e^{2x_2}$$
$$g_{12}(x_1, x_2) = g_{21}(x_1, x_2) = 2x_1e^{2x_2}$$
$$g_{22}(x_1, x_2) = 2(x_1{}^2e^{2x_2} + 1)$$

Let

$$\langle \alpha, \beta \rangle_x = \sum_{i=1}^{2} \sum_{j=1}^{2} g_{ij}(x) \langle \alpha, dx_i \rangle \langle \beta, dx_j \rangle$$

Let

$$\gamma(t) = \left(\frac{1}{t+1} \log (t+1), \log (t+1)\right)$$

for $0 \leqq t \leqq 1$. Compute the arc length along γ with respect to $\langle \alpha, \beta \rangle_x$.

11. Let $\langle \alpha, \beta \rangle_m$ be a Riemann metric on a coordinate neighborhood U of a differentiable manifold M, with coordinates $(x_1, \ldots, x_n)$. Let $\delta: M \to M$ be a diffeomorphism. Let $(y_1, \ldots, y_n)$ be coordinates in $\delta(U)$. Define

$$\langle \alpha, \beta \rangle'_y = \langle \delta_*{}^{-1}(\alpha), \delta_*{}^{-1}(\beta) \rangle_{\delta^{-1}(y)}$$

on $\delta(U)$. Show that $\langle \alpha, \beta \rangle'_y$ is a Riemann metric on $\delta(U)$. If

$$\langle \alpha, \beta \rangle_x = \sum_{i=1}^{n} \sum_{j=1}^{n} g_{ij}(x) \langle \alpha, dx_i \rangle \langle \beta, dx_j \rangle$$

and

$$\langle \alpha, \beta \rangle'_y = \sum_{i=1}^{n} \sum_{j=1}^{n} h_{ij}(y) \langle \alpha, dy_i \rangle \langle \beta, dy_j \rangle$$

find the relationship between the functions g_{ij} and h_{ij}.

12. Under the circumstances described in Exercise 11, let γ be a differentiable parametrized curve in U. Show that the arc length along $\delta\circ\gamma$ with respect to $\langle\alpha, \beta\rangle'_y$ is equal to the arc length along γ with respect to $\langle\alpha, \beta\rangle_x$.

13. Let

$$g_{11}(x_1, x_2, x_3) = \frac{x_1^2}{1 - x_1^2}$$

$$g_{22}(x_1, x_2, x_3) = \frac{x_2^2}{1 - x_2^2}$$

$$g_{33}(x_1, x_2, x_3) = 1$$

and let

$$\langle\alpha, \beta\rangle_x = \sum_{i=1}^{3} g_{ii}(x)\langle\alpha, dx_i\rangle\langle\beta, dx_j\rangle$$

on the subset $x_1^2 + x_2^2 < 1$ of E^3. Show that $\langle\alpha, \beta\rangle_x$ is a Riemann metric on this subset. Let

$$\gamma(t) = \left(\cos\frac{\pi}{6}(t + 1), \sin\frac{\pi}{6}(t + 1), t\right)$$

for $0 \leqq t \leqq 1$. Find the arc length along γ with respect to $\langle\alpha, \beta\rangle_x$.

5-5 PARACOMPACT SPACES

The topological facts in this section are needed for the proof that every differentiable manifold has a Riemann metric. This theorem is presented in the following section.

Let X be a topological space. A covering of X is called *locally finite* if every point has a neighborhood which intersects only a finite number of elements of the covering. Let $\mathfrak{C}$ and $\mathfrak{C}'$ be two coverings of X. $\mathfrak{C}'$ is called a *refinement* of $\mathfrak{C}$ if every element of $\mathfrak{C}'$ is contained in an element of $\mathfrak{C}$. A Hausdorff space X is called *paracompact* if every covering of X by open sets has a locally finite refinement which consists of open sets. It is clear that every compact Hausdorff space is paracompact.

Let X be a paracompact space. Let $\mathfrak{U} = \{U_\alpha\}$ be a covering of X by open sets. Then there is a covering $\mathfrak{W} = \{W_\beta\}$ by open sets which is a locally finite refinement of $\mathfrak{U}$. However, we may insist on more than this. We may arrange that the index sets for $\mathfrak{U}$ and $\mathfrak{W}$ be the same and that $W_\alpha \subset U_\alpha$ for each index α. To see this, let $U_{\alpha(\beta)}$ be a fixed element of $\mathfrak{U}$ which contains W_β. For each index α_0 of the index set for $\mathfrak{U}$ define

$$V_{\alpha_0} = \bigcup_{\alpha(\beta)=\alpha_0} W_\beta$$

Let $\mathfrak{V} = \{V_\alpha\}$. Then $\mathfrak{V}$ is a covering of X by open sets such that $V_\alpha \subset U_\alpha$ for every index α. Furthermore, any neighborhood which intersects only a finite number of the elements of $\mathfrak{W}$ also intersects only a finite number of the elements of $\mathfrak{V}$. Consequently $\mathfrak{V}$ is locally finite.

Theorem 5-6. *If X is a locally compact Hausdorff space with a denumerable basis, then X is paracompact.*

Proof. X has a denumerable basis $U_1, U_2, \ldots$. Since X is locally compact, we may assume that the sets $\bar{U}_1, \bar{U}_2, \ldots$ are compact. For if $x \in U_i$, then x has a neighborhood W which is contained in U_i and whose closure $\bar{W}$ is compact. There is a j such that $x \in U_j \subset W$, and it follows that $\bar{U}_j$ is compact. Consequently we may discard those among the sets $U_1, U_2, \ldots$ whose closures are not compact, and the remaining sets will still be a basis for X.

Define $W_1 = U_1$. Assuming that the open set W_{j-1} is defined so that its closure is compact, let k be the least integer such that

$$\bar{W}_{j-1} \subset U_1 \cup U_2 \cup \cdots \cup U_k$$

Then define

$$W_j = U_1 \cup \cdots \cup U_k \cup U_j$$

This procedure defines a sequence of open sets $W_1, W_2, \ldots$ whose closures are compact so that $\bar{W}_{j-1} \subset W_j$ and $\cup W_j = \cup \bar{W}_j = X$.

Let $\mathfrak{U} = \{U_\alpha\}$ be any covering of X by open sets. Let $j \geqq 2$. The compact set $\bar{W}_{j+1} - W_j$ is contained in the open set $W_{j+2} - \bar{W}_{j-1}$. $\bar{W}_{j+1} - W_j$ is covered by a finite number $U_{\alpha_1}, \ldots, U_{\alpha_r}$ of the sets of $\mathfrak{U}$. Let $V_i = U_{\alpha_i} \cap (W_{j+2} - \bar{W}_{j-1})$. Let $\mathfrak{V}_j$ denote the family $\{V_1, \ldots, V_r\}$. For $j = 1$, cover $\bar{W}_2$ with a finite number of sets of $\mathfrak{U}$ and let $\mathfrak{V}_1$ be the family of these sets. Define $\mathfrak{V} = \cup_1^\infty \mathfrak{V}_i$.

Since $\bar{W}_2 \cup \cup_2^\infty(\bar{W}_{j+1} - W_j) = X$, $\mathfrak{V}$ is a covering of X by open sets which is a refinement of $\mathfrak{U}$. If x is any point of X, there is a j such that $x \in W_j$. Clearly $\bar{W}_j$ intersects only a finite number of the sets of $\mathfrak{V}$. This proves that $\mathfrak{V}$ is locally finite and hence that $\mathfrak{U}$ has a locally finite refinement.

Since a differentiable manifold is a locally compact Hausdorff space with a denumerable basis, it follows from Theorem 5-6 that it is paracompact. However, one may prove a stronger statement.

We shall denote by B_r the set of points x in E^n with

$$|x|^2 = x_1{}^2 + \cdots + x_n{}^2 < r^2$$

Theorem 5-7. *Let M be a differentiable manifold. Let $\mathfrak{U} = \{U_\alpha\}$ be a covering of M by open sets. Then there is a denumerable family $\mathfrak{V}$ of coordinate neighborhoods of M such that the following conditions prevail:*

(i) *$\mathfrak{V}$ is a locally finite refinement of $\mathfrak{U}$.*

(ii) *If* $V \in \mathfrak{B}$ *and* φ: $V \to E^n$ *is the coordinate mapping associated with* V, *then* $\varphi(V) = B_3$.

(iii) *If for each coordinate neighborhood* V, φ *in* $\mathfrak{B}$ *we define*

$$V' = \varphi^{-1}(B_1)$$

then the family $\mathfrak{B}'$ *of the sets* V' *covers* M.

Proof. We introduce the sets $W_1, W_2, \ldots$ which were defined in the proof of Theorem 5-6. Then M is the union of the disjoint sets W_2, $W_3 - W_2$, $W_4 - W_3$, ... If $m \in M$ and, in particular, $m \in W_{j+1} - W_j$, let $\alpha(m)$ be an index such that $m \in U_{\alpha(m)}$. There is a coordinate neighborhood N_m with the coordinate mapping ν_m such that $\nu_m(m) = (0, \ldots, 0)$, $\nu_m(N_m) = B_3$, and $N_m \subset U_{\alpha(m)} \cap (W_{j+2} - \bar{W}_{j-1})$. If $m \in W_2$, we may require that $N_m \subset U_{\alpha(m)} \cap W_3$. Since the family $\mathfrak{N} = \{N_m\}$ is a refinement of $\mathfrak{U}$, it suffices to prove Theorem 5-7 for $\mathfrak{N}$ instead of $\mathfrak{U}$. Let $N'_m = \nu_m^{-1}(B_1)$ and $\mathfrak{N}' = \{N'_m\}$.

For $j \geqq 2$, $\bar{W}_{j+1} - W_j$ is covered by a finite number $N'_{m_1}, \ldots, N'_{m_r}$ of the sets of $\mathfrak{N}'$ with $m_1, \ldots, m_r \in \bar{W}_{j+1} - W_j$. Since $m_1, \ldots, m_r \in W_{j+2} - W_j$, it follows that $N_{m_1}, \ldots, N_{m_r}$ are all contained in $W_{j+3} - \bar{W}_{j-1}$. For $j = 1$, cover $\bar{W}_2$ with a finite number of the sets of $\mathfrak{N}'$. Define $V_i = N_{m_i} = \nu_{m_i}^{-1}(B_3)$. The remainder of the proof is the same as the proof of Theorem 5-6, but now the family $\mathfrak{B}$ also satisfies conditions ii and iii.

A denumerable covering $\mathfrak{B}$ of a differentiable manifold M by coordinate neighborhoods which satisfies conditions ii and iii of Theorem 5-7 will be said to be *normalized.* Theorem 5-7 then says that every covering of M by open sets has a locally finite normalized refinement.

EXERCISE

Prove that a paracompact space is normal.

5-6 THE EXISTENCE OF RIEMANN METRICS

Let M be a differentiable manifold and let f be a nonnegative differentiable function on M. The closure of the set of points m in M for which $f(m) > 0$ will be called the *carrier* of f. A family $\mathfrak{F}$ of differentiable functions on M will be called a *partition of unity* if it satisfies the following conditions:

1. Each element of $\mathfrak{F}$ is nonnegative.
2. The family of carriers of the elements of $\mathfrak{F}$ forms a locally finite

covering of M [which implies that the infinite sum $\Sigma_{f\in\mathfrak{F}} f(m)$ makes sense for each $m \in M$].

3. For each $m \in M$, $\Sigma_{f\in\mathfrak{F}} f(m) = 1$.

Lemma 5-1. *Let M be a differentiable manifold and $\mathfrak{U}$ an open covering of M. Then there is a denumerable partition of unity $\mathfrak{F} = \{f_j\}$ such that the carriers of $\mathfrak{F}$ form a refinement of $\mathfrak{U}$.*

Proof. Introduce the coverings $\mathfrak{V}$ and $\mathfrak{V}'$ of Theorem 5-7 and also the covering $\mathfrak{V}''$ consisting of the sets $V'' = \phi^{-1}(B_2)$. By Lemma 2-1, for each V_j in $\mathfrak{V}$ there is a nonnegative differentiable function g_j for which

$$g_j(m) = \begin{cases} 1 & \text{for } m \in V_j' \\ 0 & \text{for } m \notin \bar{V}_j'' \end{cases}$$

Since $\mathfrak{V}$ is a locally finite covering of M, it makes sense to define

$$g(m) = \sum_{j=1}^{\infty} g_j(m)$$

for each $m \in M$. g is then a nonzero differentiable function on M. Define

$$f_j(m) = \frac{g_j(m)}{g(m)}$$

for each V_j in $\mathfrak{V}$. Then $f_j(m)$ is nonnegative and $\Sigma_{j=1}^{\infty} f_j(m) = 1$ for each m.

Theorem 5-8. *Every differentiable manifold M can be given a Riemann metric.*

Proof. Let $\mathfrak{U}$ be a covering of M by coordinate neighborhoods. From Lemma 5-1, there is a partition of unity $\mathfrak{F} = \{f_j\}$ whose carriers form a refinement of $\mathfrak{U}$. For each j let U_j be an element of $\mathfrak{U}$ which contains the carrier of f_j. Introduce any Riemann metric $\langle\alpha, \beta\rangle_m{}^{(j)}$ into U_j. Then the infinite sum

$$\langle\alpha, \beta\rangle_m = \sum_{j=1}^{\infty} f_j(m)\langle\alpha, \beta\rangle_m{}^{(j)}$$

makes sense for each $m \in M$, and it is easy to verify that it is a Riemann metric on M.

CHAPTER 6

The Whitney Imbedding Theorem

In this chapter we shall prove a theorem due to Hassler Whitney which shows that a differentiable manifold may be looked upon as a closed submanifold of E^m if the dimension m is sufficiently large. At first hand, this result seems to indicate that our studies of abstract differentiable manifolds have been mere exercises and that we might as well have never departed from the euclidean spaces. However, such a limitation would stand in the way of a natural treatment of most differentiable manifolds and the development of intuitive insight into their properties.

6-1 SETS OF MEASURE ZERO

Let x be a point of E^n. We shall denote by $Q(x, r)$ the *hypercube* of side $r > 0$ and center x which consists of all points y such that $|y_i - x_i| \leqq \frac{1}{2}r$, $i = 1, \ldots, n$. The volume of $Q(x, r)$ is r^n. Let S be a subset of E^n. S is said to have *measure zero* if for every $\epsilon > 0$ there is a denumerable family of hypercubes which cover S and whose total volume is less than ϵ.

Let the set S have measure zero. Let $z \notin S$. Then no hypercube $Q(z, r)$ can be contained in $S \cup \{z\}$. Otherwise every covering of S by hypercubes would have total volume $\geqq r^n$. Thus every hypercube with center z contains a point not in $S \cup \{z\}$. This shows that the complement of S is dense in E^n.

Lemma 6-1. *Let U be an open subset of E^n and $\sigma: U \to E^n$ a differentiable mapping. If S is a subset of U with measure zero, then $\sigma(S)$ has measure zero.*

Proof. Let $\epsilon > 0$ be given. By assumption, S may be covered by a denumerable family of hypercubes whose total volume is less than ϵ. We shall first show that these hypercubes can be assumed to be in U. It suffices to show that, if Q is any hypercube, $Q \cap U$ may be covered by a denumerable or finite family of hypercubes contained in U whose total volume is not greater than the volume of Q.

Let $Q(x, r)$ be a hypercube which is not contained in U. Subdivide $Q(x, r)$ into 2^n hypercubes of side $\frac{1}{2}r$. These will overlap only on their boundary hyperplanes. Discard those small hypercubes that are disjoint from U, retain those that are contained in U, and subdivide similarly the remaining ones. Again discard the smallest hypercubes that are disjoint from U, retain those contained in U, and subdivide the remaining ones. Repeat the process so as to obtain a finite or denumerable family of hypercubes that are contained in U and whose total volume does not exceed the volume of $Q(x, r)$.

Let $y \in Q(x, r) \cap U$. Then y has a neighborhood of radius $\rho > 0$ which is contained in U. When the subdivision process has proceeded to the stage where the sides of the hypercubes are less than $\frac{1}{2}\rho$, then y will certainly have to be contained in a hypercube that was retained. Thus the denumerable or finite family which we have constructed covers $Q(x, r) \cap U$.

Let the mapping σ be given by $\sigma(x) = (f_1(x), \ldots, f_n(x))$. Let Q be any hypercube which is contained in U. Let $K(Q)$ denote the maximum of all the functions $|\partial f_i/\partial x_j|$ on the compact set Q. Then for any two points x and y in Q we have

$$f_i(x) - f_i(y) = \sum_{j=1}^{n} [f_i(y_i, \ldots, y_{j-1}, x_j, \ldots, x_n) - f_i(y_1, \ldots, y_j, x_{j+1}, \ldots, x_n)]$$

If we apply the theorem of the mean to each of the terms of the sum on the right, then

$$f_i(x) - f_i(y) = \sum_{j=1}^{n} \left.\frac{\partial f_i}{\partial x_j}\right|_{z^i} (x_j - y_j)$$

where $z^1, \ldots, z^n$ are points in Q. Hence

$$|f_i(x) - f_i(y)| \leqq K(Q) \sum_{j=1}^{n} |x_j - y_j|$$

for $i = 1, \ldots, n$. It follows that if $Q(x, r)$ is contained in U then

$$\sigma(Q(x, r)) \subset Q(\sigma(x), nK(Q(x, r))r)$$

Let $\epsilon > 0$ be given. Cover S by hypercubes $Q_1, Q_2, \ldots$ contained in U. For each i there is a hypercube $\bar{H}_i$ contained in U whose interior H_i satisfies $Q_i \subset H_i$. Then $S \cap H_i$ has measure zero. Cover $S \cap H_i$ with hypercubes $Q(y_{i1}, r_1), Q(y_{i2}, r_2), \ldots$ which are contained in H_i and whose total volume is less than $\epsilon/2^i n^n K(\bar{H}_i)^n$. Since

$$\sigma(Q(y_{ij}, r_j)) \subset Q(\sigma(y_{ij}), nK(\bar{H}_i)r_j)$$

we find that $\sigma(S \cap H_i)$ is covered by the hypercubes $Q(\sigma(y_{i1}), nK(\bar{H}_i)r_1)$, $Q(\sigma(y_{i2}), nK(\bar{H}_i)r_2)$, . . . whose total volume is less than $\epsilon/2^i$. It follows that $\sigma(S)$ is covered by the denumerable family $\{Q(\sigma(y_{ij}), nK(\bar{H}_i)r_j)\}$ of hypercubes whose total volume is less than $\epsilon\Sigma_{i=1}^{\infty}(\frac{1}{2})^i = \epsilon$. This shows that $\sigma(S)$ is of measure zero.

Lemma 6-2. *Let U be an open subset of E^n and $\sigma: U \to E^p$ a differentiable mapping. If $n < p$, then $\sigma(U)$ has measure zero.*

Proof. Let $z = (z_1, \ldots, z_{p-n})$ be a point in E^{p-n}. We shall show that $U \times \{z\}$ has measure zero in E^p. It suffices to show that $E^{p-1} \times \{z_{p-n}\}$ has measure zero since $U \times \{z\}$ is a subset of this set. Let $\delta > 0$, and consider the region H_δ in E^p between the hypersurfaces

$$x_p = z_{p-n} \pm \frac{\delta}{1 + (x_1{}^2 + \cdots + x_{p-1}^2)^{p-1}}$$

H_δ contains the set $E^{p-1} \times \{z_{p-n}\}$, its volume $v(H_\delta)$ is finite, and $v(H_\delta) = \delta v(H_1)$. Clearly it is possible to fit a sequence of hypercubes into H_δ in such a way that the total volume of these hypercubes is less than $v(H_\delta)$ and yet they cover $E^{p-1} \times \{z_{p-n}\}$. Since $v(H_\delta)$ may be made arbitrarily small, the set $E^{p-1} \times \{z_{p-n}\}$ has measure zero.

Let $\pi: E^p = E^n \times E^{p-n} \to E^n$ be the natural projection. Then $\sigma\circ\pi: U \times E^{p-n} \to E^p$ is a differentiable mapping. By Lemma 6-1, $\sigma\circ\pi(U \times \{z\}) = \sigma(U)$ has measure zero.

Let M be a differentiable manifold. Let S be a subset of M. Then S is said to have measure zero if for every coordinate neighborhood U, φ of M the set $\varphi(S \cap U)$ has measure zero. It follows immediately from Lemma 6-2 that *if σ is a differentiable mapping of a differentiable manifold M into a differentiable manifold M' and if the dimension of M is less than the dimension of M', then $\sigma(M)$ has measure zero.*

By refining the methods of this section, it is possible to obtain the following statement which is due to A. Sard. Let U be an open subset of E^n and $\sigma: U \to E^p$ a differentiable mapping. Let σ_* be the induced linear transformation on the tangent space to E^n. Let S_r be the set of points of E^n where σ_* is of rank r. If $r < p$, then $\sigma(S_r)$ is of measure zero. We shall not give the proof.

EXERCISE

Compute the volume of H_1.

6-2 APPROXIMATION BY REGULAR MAPPINGS

Let $\mathfrak{M}(p \times n)$ be the vector space of all $p \times n$ matrices. $\mathfrak{M}(p \times n)$ is isomorphic to V^{pn} under the mapping

$$\begin{pmatrix} a_{11} & \cdots & a_{1n} \\ \cdots & \cdots & \cdots \\ a_{p1} & \cdots & a_{pn} \end{pmatrix} \rightarrow (a_{11}, \ldots, a_{1n}, a_{21}, \ldots, a_{pn})$$

and so we may give to $\mathfrak{M}(p \times n)$ the differentiable structure of V^{pn}. We denote by $\mathfrak{M}(p \times n, k)$ the subset of $\mathfrak{M}(p \times n)$ which consists of the matrices of rank k.

Lemma 6-3. *If $k \leqq \min \{p, n\}$, then $\mathfrak{M}(p \times n, k)$ is a submanifold of $\mathfrak{M}(p \times n)$ of dimension $k(p + n - k)$.*

Proof. Let $\mathcal{P}$ be a nonsingular $p \times p$ matrix and $\mathcal{Q}$ a nonsingular $n \times n$ matrix. Then the mapping $\mathfrak{M} \rightarrow \mathcal{P}\mathfrak{M}\mathcal{Q}$ of $\mathfrak{M}(p \times n)$ onto $\mathfrak{M}(p \times n)$ is a diffeomorphism. If $\mathfrak{M}$ is any element of $\mathfrak{M}(p \times n, k)$, there are nonsingular matrices $\mathcal{P}$ and $\mathcal{Q}$ such that $\mathcal{P}\mathfrak{M}\mathcal{Q}$ has the special form

$$\begin{pmatrix} \mathcal{A} & \mathcal{B} \\ \mathcal{C} & \mathcal{D} \end{pmatrix}$$

with $\mathcal{A}$ a nonsingular $k \times k$ matrix. Consequently it suffices to produce coordinate neighborhoods in $\mathfrak{M}(p \times n, k)$ for matrices of this form.

Let

$$\mathfrak{M}_0 = \begin{pmatrix} \mathcal{A}_0 & \mathcal{B}_0 \\ \mathcal{C}_0 & \mathcal{D}_0 \end{pmatrix}$$

where $\mathfrak{M}_0 \in \mathfrak{M}(p \times n, k)$ and $\mathcal{A}_0$ is a nonsingular $k \times k$ matrix. Let $\mathcal{A} \in \mathfrak{M}(k \times k)$, and denote by $s(\mathcal{A}, \mathcal{A}_0)$ the maximum of the absolute values of the components of the matrix $\mathcal{A} - \mathcal{A}_0$. Since $\det \mathcal{A}_0 \neq 0$ and $\det \mathcal{A}$ is a continuous function of the components of $\mathcal{A}$, it follows that there is a $\delta > 0$ such that if $s(\mathcal{A}, \mathcal{A}_0) < \delta$ then $\det \mathcal{A} \neq 0$, and so $\mathcal{A}$ is nonsingular. Let U consist of all matrices of $\mathfrak{M}(p \times n)$ of the form

$$\mathfrak{M} = \begin{pmatrix} \mathcal{A} & \mathcal{B} \\ \mathcal{C} & \mathcal{D} \end{pmatrix}$$

where $\mathcal{A} \in \mathfrak{M}(k \times k)$ and $s(\mathcal{A}, \mathcal{A}_0) < \delta$. Then U is an open subset of $\mathfrak{M}(p \times n)$.

The matrix

$$\begin{pmatrix} \mathcal{E}_k & 0 \\ -\mathcal{C}\mathcal{A}^{-1} & \mathcal{E}_{p-k} \end{pmatrix} \begin{pmatrix} \mathcal{A} & \mathcal{B} \\ \mathcal{C} & \mathcal{D} \end{pmatrix} = \begin{pmatrix} \mathcal{A} & \mathcal{B} \\ 0 & -\mathcal{C}\mathcal{A}^{-1}\mathcal{B} + \mathcal{D} \end{pmatrix}$$

where $\mathcal{E}_j$ is the $j \times j$ identity matrix, has the same rank as the matrix

$$\mathfrak{M} = \begin{pmatrix} \mathcal{A} & \mathcal{B} \\ \mathcal{C} & \mathcal{D} \end{pmatrix}$$

Consequently the matrix $\mathfrak{M}$ has rank k if and only if $\mathcal{D} = \mathcal{C}\mathcal{A}^{-1}\mathcal{B}$. It follows that the matrices of the open set $U \cap \mathfrak{M}(p \times n,\ k)$ in $\mathfrak{M}(p \times n,\ k)$ are precisely the matrices of the form

$$\begin{pmatrix} \mathcal{A} & \mathcal{B} \\ \mathcal{C} & \mathcal{C}\mathcal{A}^{-1}\mathcal{B} \end{pmatrix}$$

where $\mathcal{A} \in \mathfrak{M}(k \times k)$ and $s(\mathcal{A},\ \mathcal{A}_0) < \delta$. This open set contains the matrix $\mathfrak{M}_0$.

The set of all matrices in $\mathfrak{M}(p \times n)$ of the form

$$\begin{pmatrix} \mathcal{A} & \mathcal{B} \\ \mathcal{C} & 0 \end{pmatrix}$$

where $\mathcal{A} \in \mathfrak{M}(k \times k)$, has the differentiable structure of euclidean space of dimension $pn - (p - k)(n - k) = k(p + n - k)$. Let W be the subset of these matrices for which $s(\mathcal{A},\ \mathcal{A}_0) < \delta$. Then W is an open subset of $E^{k(p+n-k)}$. Consider the one-to-one mapping $\sigma\colon W \to \mathfrak{M}(p \times n)$, which is onto $U \cap \mathfrak{M}(p \times n,\ k)$, defined by

$$\sigma\begin{pmatrix} \mathcal{A} & \mathcal{B} \\ \mathcal{C} & 0 \end{pmatrix} = \begin{pmatrix} \mathcal{A} & \mathcal{B} \\ \mathcal{C} & \mathcal{C}\mathcal{A}^{-1}\mathcal{B} \end{pmatrix}$$

σ is clearly differentiable. If $\tau\colon U \to W$ is defined by

$$\tau\begin{pmatrix} \mathcal{A} & \mathcal{B} \\ \mathcal{C} & \mathcal{D} \end{pmatrix} = \begin{pmatrix} \mathcal{A} & \mathcal{B} \\ \mathcal{C} & 0 \end{pmatrix}$$

then τ is a differentiable mapping and $\tau \circ \sigma =$ identity on W. It follows that σ is regular on W; hence $U \cap \mathfrak{M}(p \times n,\ k)$ is a submanifold of $\mathfrak{M}(p \times n)$. If we take $U \cap \mathfrak{M}(p \times n,\ k)$ to be a coordinate neighborhood of $\mathfrak{M}(p \times n,\ k)$, then $\mathfrak{M}(p \times n,\ k)$ becomes a submanifold of $\mathfrak{M}(p \times n)$ of dimension $k(p + n - k)$.

Lemma 6-4. *Let U be an open subset of V^n, and let $\sigma\colon U \to V^p$, $p \geqq 2n$, be a differentiable mapping. For each $\epsilon > 0$ there is a $p \times n$ matrix $\mathcal{A} = (a_{ij})$ such that $|a_{ij}| < \epsilon$ for $i = 1, \ldots, p$; $j = 1, \ldots, n$ and $\tau\colon U \to E^p$ defined by*

$$\tau(x) = \sigma(x) + \mathcal{A} \cdot x$$

is regular on U.

Proof. Define the mapping $\rho_k\colon \mathfrak{M}(p \times n,\ k) \times U \to \mathfrak{M}(p \times n)$ by

$$\rho_k(\mathfrak{M},\ x) = \mathfrak{M} - \mathcal{J}(x)$$

where $\mathcal{J}(x)$ is the jacobian matrix of the mapping σ, defined as

$$\mathcal{J}(x) = \begin{pmatrix} \frac{\partial y_1}{\partial x_1} & \cdots & \frac{\partial y_1}{\partial x_n} \\ \cdots & \cdots & \cdots \\ \frac{\partial y_p}{\partial x_1} & \cdots & \frac{\partial y_p}{\partial x_n} \end{pmatrix}$$

where $y = \sigma(x)$. ρ_k is a differentiable mapping of a differentiable manifold of dimension $k(p + n - k) + n$ into a differentiable manifold of dimension pn. Let $k \leqq n - 1$. Since $p \geqq 2n$, we have

$$\begin{aligned} pn - [k(p + n - k) + n] &= (p - k)(n - k) - n \\ &\geqq (2n - n + 1)(n - n + 1) - n = 1 > 0 \end{aligned}$$

From Lemma 6-2 we know that the image domain of ρ_k has measure zero in $\mathfrak{M}(p \times n)$ for $k \leqq n - 1$. Hence the complement of the image domain of ρ_k is dense in $\mathfrak{M}(p \times n)$.

Let $\epsilon > 0$. There must be a matrix $\mathfrak{A} = (a_{ij})$ with $|a_{ij}| < \epsilon$ for all i and j which is not in the image domain of $\rho_0, \rho_1, \ldots, \text{or } \rho_{n-1}$. Let $\tau(x) = \sigma(x) + \mathfrak{A} \cdot x$ for $x \in U$. Then the jacobian matrix of the mapping τ is $\mathcal{J}(x) + \mathfrak{A}$. This jacobian matrix is not in $\mathfrak{M}(p \times n, k)$ for $k \leqq n - 1$, since otherwise $\rho_k(\mathcal{J}(x) + \mathfrak{A}, x) = \mathfrak{A}$ would be in the image domain of ρ_k. Hence $\mathcal{J}(x) + \mathfrak{A}$ has rank n for every x in U. This shows that τ is regular on U.

If x and y are two points of V^p, we shall denote by $d(x, y)$ the distance between x and y when V^p is identified with E^p.

Theorem 6-1. *Let M be a differentiable manifold of dimension n, and let C be a closed subset of M. Let $\sigma\colon M \to V^p$, $p \geqq 2n$, be a differentiable mapping which is regular on C. For each continuous positive function η on M there is a regular differentiable mapping $\tau\colon M \to V^p$ such that $\tau(m) = \sigma(m)$ for $m \in C$ and $d(\tau(m), \sigma(m)) < \eta(m)$ for $m \in M$.*

Proof. If m is a point of C, there is a neighborhood of m on which σ is regular. It follows that C is contained in an open set A on which σ is regular. The open sets $\{A, \text{complement of } C\}$ cover M. By Theorem 5-7, this covering has a denumerable, locally finite, normalized refinement $\mathfrak{B}$. We shall number the sets V_j of $\mathfrak{B}$ with positive and negative integers so that $V_j \subset A$ if $j < 0$ and $V_j \subset$ complement of C if $j > 0$. If φ_j is the coordinate mapping of the coordinate neighborhood V_j, let $V_j' = \varphi_j^{-1}(B_2)$ and $V_j'' = \varphi_j^{-1}(B_1)$, where B_r consists of the points x in V^n for which $d(x, 0) < r$. Recall that the sets $\{V_j''\}$ cover M.

We are going to construct a sequence $\sigma_0, \sigma_1, \ldots$ of differentiable mappings $\sigma_k \colon M \to V^p$ such that the following conditions prevail:

1. $\sigma_0 = \sigma$.
2. σ_k is regular on $F_k = \overline{\cup_{j<0} V_j''} \cup \overline{V_1''} \cup \cdots \cup \overline{V_k''}$ for $k = 1, 2, \ldots$.
3. $\sigma_k(m) = \sigma_{k-1}(m)$ for $m \notin V_k'$, $k = 1, 2, \ldots$.
4. $d(\sigma_k(m), \sigma_{k-1}(m)) < \eta(m)/2^k$ for $m \in M$, $k = 1, 2, \ldots$.

Suppose that $\sigma_0, \sigma_1, \ldots, \sigma_{k-1}$ have already been constructed. Let ψ be a differentiable function on V^n such that $0 \leqq \psi(x) \leqq 1$, $\psi(x) = 1$ for $x \in B_1$, and $\psi(x) = 0$ for $x \notin B_2$. For any $p \times n$ matrix $\mathfrak{M}$ define

$$\Phi_{\mathfrak{M}}(x) = \sigma_{k-1} \circ \varphi_k^{-1}(x) + \psi(x)\mathfrak{M} \cdot x$$

for $x \in B_3$, where φ_k is the coordinate mapping of V_k. Then $\Phi_{\mathfrak{M}} \colon B_3 \to V^p$ is a differentiable mapping.

Let $\mathcal{J}(\mathfrak{M}, x)$ be the jacobian matrix of $\Phi_{\mathfrak{M}}$ and $\mathcal{J}_k(x)$ the jacobian matrix of $\sigma_{k-1} \circ \varphi_k^{-1}$. If $\mathfrak{M} = (m_{ij})$, then

$$\mathcal{J}(\mathfrak{M}, x) = \mathcal{J}_k(x) + \psi(x)\mathfrak{M} + \left(\frac{\partial \psi}{\partial x_j} \sum_{k=1}^{m} m_{ik} x_k \right)$$

Consequently the components of $\mathcal{J}(\mathfrak{M}, x)$ are differentiable functions of the components of $\mathfrak{M}$ and x. We may regard the mapping

$$(\mathfrak{M}, x) \to \mathcal{J}(\mathfrak{M}, x)$$

as a mapping from $V^{pn} \times B_3$, which is an open subset of V^{pn+n}, into V^{pn}.

Let $z \in \varphi_k(\overline{V'} \cap F_{k-1})$. Then $\mathcal{J}(0, z) = \mathcal{J}_k(z)$ is of rank n since $\sigma_{k-1} \circ \varphi_k^{-1}$ is regular at z. Hence there is a neighborhood of $(0, z)$ in $V^{pn} \times B_3$, of the form $W_z \times U_z$ with $0 \in W_z \subset V^{pn}$ and $z \in U_z \subset B_3$, on which $\mathcal{J}(\mathfrak{M}, x)$ is of rank n. The set $\{0\} \times \varphi_k(\overline{V'} \cap F_{k-1})$ is compact, and so it is covered by a finite number of these neighborhoods. From this we may conclude that there is a neighborhood W of 0 in V^{pn} such that, if $\mathfrak{M} \in W$ and $z \in \varphi_k(\overline{V'} \cap F_{k-1})$, then $\mathcal{J}(\mathfrak{M}, z)$ has rank n and so $\Phi_{\mathfrak{M}}$ is regular at z. We may also assume that W is sufficiently small so that $d(0, \mathfrak{M} \cdot x) < \eta_k/2^k$ for $x \in \bar{B}_2$ and $\mathfrak{M} \in W$, where η_k is the minimum of the function η on the compact set $\overline{V_k'}$.

Now apply Lemma 6-4 with U taken to be B_2, σ taken to be $\sigma_{k-1} \circ \varphi_k^{-1}$, and ϵ determined by the size of the neighborhood W. Then there is an $\mathfrak{M}$ in W such that the mapping

$$x \to \sigma_{k-1} \circ \varphi_k^{-1}(x) + \mathfrak{M} \cdot x$$

is regular on B_2. Using this $\mathfrak{M}$, define $\sigma_k \colon M \to V^p$ by

$$\sigma_k(m) = \begin{cases} \sigma_{k-1}(m) + \psi(\varphi_k(m))\mathfrak{M} \cdot \varphi_k(m) & \text{for } m \in V_k \\ \sigma_{k-1}(m) & \text{for } m \notin V_k' \end{cases}$$

σ_k is pieced together from two differentiable mappings which agree on the open set $V_k - \overline{V'_k}$, and so it is differentiable on M. Since $\psi(\varphi_k(m)) = 1$ for $m \in \overline{V''_k}$, our choice of $\mathfrak{M}$ guarantees that σ_k is regular for $m \in \overline{V''_k}$. Since $\mathfrak{M}$ is in W, σ_k is regular on $V'_k \cap F_{k-1}$. On the remainder of the set F_{k-1}, $\sigma_k(m) = \sigma_{k-1}(m)$, and so σ_k is regular. Thus σ_k is regular on all of F_{k-1} and hence on F_k. Finally we observe that

$$d(\sigma_k(m), \sigma_{k-1}(m)) = \psi(\varphi_k(m))\, d(0, \mathfrak{M} \cdot \varphi_k(m)) < \frac{\eta_k}{2^k} \leqq \frac{\eta(m)}{2^k}$$

for $m \in V'_k$. Since this distance is 0 for $m \notin V'_k$, we find that $d(\sigma_k(m), \sigma_{k-1}(m)) < \eta(m)/2^k$ for all $m \in M$. Thus we have completed the construction of the mappings $\sigma_0, \sigma_1, \ldots$.

Let m° be a point of M. There is a neighborhood N of m° which intersects only a finite number of the sets $V_1, V_2, \ldots$. Thus for sufficiently large k the mappings $\sigma_k, \sigma_{k+1}, \ldots$ are the same on N. We shall define $\tau(m^\circ) = \sigma_k(m^\circ)$. Since $\tau(m) = \sigma_k(m)$ for all $m \in N$, we see that $\tau\colon M \to V^p$ is differentiable. It follows from our construction that τ is regular on $\bigcup_{k=1}^{\infty} F_k = M$.

Let $m \in C$. If $m \in V'_k$, then V_k is not contained in the complement of C and so $k < 0$. Thus $m \notin V'_k$ for all $k > 0$, and it follows that $\tau(m) = \sigma_0(m) = \sigma(m)$. Finally we observe that

$$d(\tau(m), \sigma(m)) \leqq d(\tau(m), \sigma_k(m)) + \Sigma_{j=1}^{k} d(\sigma_j(m), \sigma_{j-1}(m))$$

which for k sufficiently large gives

$$d(\tau(m), \sigma(m)) < \eta(m) \sum_{j=1}^{k} (1/2)^j < \eta(m)$$

for each m in M.

6-3 APPROXIMATION BY ONE-TO-ONE MAPPINGS

Let $\sigma\colon M \to N$ be a regular differentiable mapping of one differentiable manifold into another. If σ is one-to-one, we know that $\sigma(M)$ is a submanifold of N. In this case we shall call σ a *diffeomorphism* of M into N.

Theorem 6-2. *Let M be a differentiable manifold of dimension n and $\sigma\colon M \to V^p$, $p \geqq 2n + 1$, a regular differentiable mapping. Let σ be one-to-one on an open set U which contains the closed set C. For each positive continuous function η on M there is a diffeomorphism $\tau\colon M \to V^p$ such that $\tau(m) = \sigma(m)$ for $m \in C$ and $d(\tau(m), \sigma(m)) < \eta(m)$ for $m \in M$.*

Proof. The proof of this theorem proceeds in much the same manner as the proof of Theorem 6-1. For this reason many of the details will be left to the reader.

Let $m \in M$. Since σ is regular at m, there is a neighborhood U_m of m on which σ is one-to-one. We may assume that U_m is either in U or in the complement of C. Let $\mathfrak{V} = \{V_j\}$ be a locally finite, normalized refinement of the covering $\{U_m\}$, and define V'_j and V''_j as in Theorem 6-1. We shall renumber the sets of $\mathfrak{V}$ with positive and negative integers so that $V_j \subset U$ if $j < 0$ and $V_j \subset$ complement of C if $j > 0$.

We shall construct a sequence $\sigma_0, \sigma_1, \ldots$ of regular differentiable mappings $\sigma_k \colon M \to V^p$ such that the following conditions prevail:

1. $\sigma_0 = \sigma$.
2. $\sigma_k(m) = \sigma_{k-1}(m)$ for $m \notin V_k$, $k = 1, 2, \ldots$.
3. $\sigma_k(m) = \sigma_k(m')$ only if $\sigma_{k-1}(m) = \sigma_{k-1}(m')$ for $m, m' \in M$.
4. $d(\sigma_k(m), \sigma_{k-1}(m)) < \eta(m)/2^k$ for $m \in M$, $k = 1, 2, \ldots$.

Suppose that $\sigma_0, \ldots, \sigma_{k-1}$ have been constructed. Let ψ be the function from the proof of Theorem 6-1. If $z \in V^p$, define

$$\Phi_z(x) = \sigma_{k-1} \circ \varphi_k^{-1}(x) + \psi(x)z$$

for $x \in B_3$. Then there is a neighborhood W of 0 in V^p such that, if $z \in W$ and $x \in B_2$, then Φ_z is regular at x. The neighborhood W may be chosen so small that for $z \in W$ we have $d(0, z) < \eta_k/2^k$, where η_k is the minimum of η on the compact set $\overline{V}_k$.

Define

$$\theta_k(m) = \begin{cases} \psi(\varphi_k(m)) & \text{for } m \in V_k \\ 0 & \text{for } m \notin V_k \end{cases}$$

The function θ_k is differentiable on M. Let Q_k be the open subset of $M \times M$ consisting of all points (m, m') such that $\theta_k(m) \neq \theta_k(m')$. Define

$$\rho_k(m, m') = \frac{-1}{\theta_k(m) - \theta_k(m')} [\sigma_{k-1}(m) - \sigma_{k-1}(m')]$$

for $(m, m') \in Q_k$. Then $\rho_k \colon Q_k \to V^p$ is a differentiable mapping. Since $p > 2n$, the image set $\rho_k(Q_k)$ has measure zero. The complement of $\rho_k(Q_k)$ is therefore dense in V^p, and it is possible to find a z in W such that $z \notin \rho_k(Q_k)$ and $z \neq 0$. Using this z, define $\sigma_k \colon M \to V^p$ by $\sigma_k(m) = \sigma_{k-1}(m) + \theta_k(m)z$. σ_k is differentiable and regular on M since $\sigma_k(m) = \Phi_z(\varphi_k(m))$ for $m \in V'_k$ and $\sigma_k(m) = \sigma_{k-1}(m)$ for $m \notin V'_k$. Furthermore, if $\sigma_k(m) = \sigma_k(m')$, then

$$\sigma_{k-1}(m) - \sigma_{k-1}(m') = -[\theta_k(m) - \theta_k(m')]z$$

Since $z \notin \rho_k(Q_k)$, it follows that $\theta_k(m) = \theta_k(m')$ and hence that $\sigma_{k-1}(m) = \sigma_{k-1}(m')$. Finally,

$$d(\sigma_k(m), \sigma_{k-1}(m)) = \theta_k(m)d(0, z) < \frac{\eta(m)}{2^k}$$

for $m \in V_k$ and hence on all of M. This completes the construction of the mappings $\sigma_0, \sigma_1, \ldots$.

The mapping $\tau\colon M \to E^p$ is defined as in the proof of Theorem 6-1. As in that theorem, τ is a regular differentiable mapping on all of M which is equal to σ on C and which satisfies $d(\tau(m), \sigma(m)) < \eta(m)$ for $m \in M$. It remains to prove that τ is one-to-one. Suppose that $\tau(m) = \tau(m')$. Then there is a j such that

$$\begin{aligned} \tau(m) &= \sigma_j(m) = \sigma_{j+1}(m) = \cdots \\ \tau(m') &= \sigma_j(m') = \sigma_{j+1}(m') = \cdots \end{aligned}$$

Hence $\sigma_k(m) = \sigma_k(m')$ for $k \geqq j$, and it follows that $\sigma_k(m) = \sigma_k(m')$ for all $k \geqq 0$. In particular, $\sigma(m) = \sigma(m')$. Now σ is one-to-one on each of the sets V_k. Let $m \in V_k''$. Then $\sigma_k(m) = \sigma_{k-1}(m) + z$. If $m' \notin V_k$, then $\sigma_k(m) = \sigma_k(m') = \sigma_{k-1}(m') = \sigma_{k-1}(m)$, which implies that $z = 0$. Since this is not true, we conclude that $m' \in V_k$, and so $m' = m$, which completes the proof.

6-4 THE IMBEDDING THEOREM

Let M be a differentiable manifold. Let $m^1, m^2, \ldots$ be a sequence of points of M. This sequence will be said to be *convergent* if there is a coordinate neighborhood U, φ and an integer k such that $m^j \in U$ for $j \geqq k$ and the sequence $\varphi(m^k), \varphi(m^{k+1}), \ldots$ is convergent. Let x° be the limit of the sequence $\varphi(m^k), \varphi(m^{k+1}), \ldots$. It is easy to see that the point $\varphi^{-1}(x^\circ)$ does not depend upon which coordinate neighborhood we use, and we shall call $m^\circ = \varphi^{-1}(x^\circ)$ the *limit* of the sequence $m^1, m^2, \ldots$. More generally a point m' will be called a *limit point* of a sequence $m^1, m^2, \ldots$ if m' is the limit of a subsequence of $m^1, m^2, \ldots$.

Let $\sigma\colon M \to E^p$ be a continuous mapping. Consider the set of all sequences $\{m^i\}$ of points of M which do not have any limit points. The corresponding sequences $\{\sigma(m^i)\}$ of points of E^p will generally not converge, but some of them may. Let $L(\sigma)$ denote the set of the limits of those sequences $\{\sigma(m^i)\}$ which do converge. The reader may easily prove that $\sigma(M)$ is a closed subset of E^p if and only if $L(\sigma) \subset \sigma(M)$.

Lemma 6-5. *If M is a differentiable manifold, then there is a differentiable function $\sigma: M \to E^1$ with $L(\sigma)$ empty.*

Proof. Let $\{V_j, \varphi_j\}$, $j = 1, 2, \ldots$, be a denumerable, locally finite, normalized covering of M by coordinate neighborhoods. Let V'_j, V''_j, and ψ be defined as in the proof of Theorem 6-1. If $m^\circ \in M$, there is a k such that $m^\circ \notin V_j$ for $j \geqq k$, and we may define

$$\sigma(m^\circ) = \sum_{j=1}^{k} j\psi(\varphi_j(m^\circ)) = \sum_{j=1}^{\infty} j\psi(\varphi_j(m^\circ))$$

Since this same integer k works for all the points of some neighborhood of m°, the mapping $\sigma: M \to E^1$ is differentiable.

Let $m^1, m^2, \ldots$ be a sequence of points of M which does not have any limit points. Let q be a positive integer, and consider the compact set $\overline{V'_1} \cup \cdots \cup \overline{V'_q}$. Then only a finite number of the points $m^1, m^2, \ldots$ may lie in $\overline{V'_1} \cup \cdots \cup \overline{V'_q}$. Thus there is an integer j such that $m^j \notin \overline{V'_1} \cup \cdots \cup \overline{V'_q}$. Then $\psi(\varphi_k(m^j)) = 0$ for $k = 1, \ldots, q$, and so $\sigma(m^j) > q$. The sequence $\sigma(m^1), \sigma(m^2), \ldots$ is therefore unbounded and cannot converge. This proves that $L(\sigma)$ is empty.

Theorem 6-3 (Whitney Imbedding Theorem). *If M is a differentiable manifold of dimension n, then there is a diffeomorphism $\tau: M \to E^{2n+1}$ such that $\tau(M)$ is closed in E^{2n+1}.*

Proof. By Lemma 6-5, there is a differentiable function σ on M with $L(\sigma)$ empty. Define $\sigma_1: M \to E^{2n+1}$ by $\sigma_1(m) = (\sigma(m), 0, \ldots, 0)$. Then σ_1 is a differentiable mapping with $L(\sigma_1)$ empty. By Theorem 6-1, there is a regular differentiable mapping $\sigma_2: M \to E^{2n+1}$ such that $d(\sigma_2(m), \sigma_1(m)) < \frac{1}{2}$ for $m \in M$. By Theorem 6-2, there is a diffeomorphism $\tau: M \to E^{2n+1}$ such that $d(\tau(m), \sigma_2(m)) < \frac{1}{2}$ for $m \in M$. Hence $d(\tau(m), \sigma_1(m)) < 1$ for $m \in M$. We shall prove that $L(\tau)$ is empty, from which we may conclude that $\tau(M)$ is a closed subset of E^{2n+1}.

Let $x \in L(\tau)$. Then there is a sequence $m^1, m^2, \ldots$ of points of M such that $x = \lim_{j\to\infty}\tau(m^j)$ but $m^1, m^2, \ldots$ does not have any limit points. But

$$\begin{aligned} d(x, \sigma_1(m^j)) &\leqq d(x, \tau(m^j)) + d(\tau(m^j), \sigma_1(m^j)) \\ &\leqq d(x, \tau(m^j)) + 1 \end{aligned}$$

Consequently the sequence $\sigma_1(m^1), \sigma_1(m^2), \ldots$ is bounded. A subsequence $\sigma_1(m^{i_1}), \sigma_1(m^{i_2}), \ldots$ must therefore converge. Since the sequence $m^{i_1}, m^{i_2}, \ldots$ does not have any limit points, this implies that $L(\sigma_1)$ is not empty. Since this is not true, $L(\tau)$ must be empty.

CHAPTER 7

Lie Groups and Their One-parameter Subgroups

The previous chapters of this book have been devoted primarily to the formulation of the definition of a differentiable manifold and the illustration of this definition by examples. In the remainder of the book we shall investigate some of the consequences which the existence of a differentiable structure entails in various situations.

7-1 LIE GROUPS

Roughly speaking, a Lie group is a differentiable manifold which is also a group in such a way that the group structure is compatible with the differentiable structure. We shall give the precise definition.

Let $f(x_1, \ldots, x_n)$ be a real-valued function defined on an open subset U of E^n. f will be said to be *real-analytic* at the point z in U if there is a neighborhood of z on which f can be represented by means of its Taylor expansion in the variables $x_1, \ldots, x_n$. f is real-analytic on U if it is real-analytic at each point of U. From the theory of power series, we know that such a function has partial derivatives of all orders and is therefore differentiable. The following example shows that even for functions of one variable a differentiable function need not be real-analytic. Let

$$f(t) = \begin{cases} e^{-1/t^2} & \text{for } t \neq 0 \\ 0 & \text{for } t = 0 \end{cases}$$

Then f is differentiable even at the point $t = 0$. But all derivatives of f are equal to zero at $t = 0$. The only real-analytic function with this property is the function which is identically zero. Hence f cannot be real-analytic at $t = 0$.

A mapping of an open subset of E^n into E^n is called real-analytic if the real-valued functions which describe this mapping are real-analytic. A differentiable manifold is called real-analytic if it may be covered by a family of coordinate neighborhoods among which the coordinate transformations are real-analytic mappings. It follows that the product of two real-analytic manifolds is real-analytic and that a submanifold of a real-analytic manifold is real-analytic.

It is necessary to caution the reader that two or more distinct real-analytic structures may arise from the same differentiable structure on a manifold. For example, the mappings

$$\varphi_1(t) = \int_0^t (1 + e^{-1/s^2})\, ds$$

and $\varphi_2(t) = t$ are coordinate mappings of E^1 which give rise to incompatible real-analytic structures on E^1.

Let M be a real-analytic manifold. Let f be a function on an open subset U of M. If $m \in U$, then f is said to be real-analytic at m if there is a coordinate neighborhood V, φ (of the real-analytic structure) of the point m such that $f \circ \varphi^{-1}$ is real-analytic at $\varphi(m)$. f is real-analytic on U if it is real-analytic at each point of U. If $\sigma: N \to M$ is a mapping of one real-analytic manifold into another, σ is said to be real-analytic if for every real-analytic function f on an open subset U of M the function $f \circ \sigma$ is real-analytic on $\sigma^{-1}(U)$.

Let X be a set of points. We shall assume that X is a real-analytic manifold and also that X is a group. When we speak of X as a manifold, we shall denote it by M, and when we speak of X as a group we shall denote it by G, at least for the present. We shall denote the product of two elements g_1 and g_2 of G by g_1g_2. Form the product manifold $M \times M$. If g_1 and g_2 are elements of G (and hence of M), then (g_1, g_2) is an element of $M \times M$, and we may consider the mapping μ_1: $M \times M \to M$ defined by $\mu_1(g_1, g_2) = g_1g_2$ and the mapping μ_2: $M \to M$ defined by $\mu_2(g) = g^{-1}$. If the mappings μ_1 and μ_2 are real-analytic, then X together with the structures of M and of G is called a *Lie group*.

The reader is already familiar with a number of Lie groups such as the additive group of V^n, the multiplicative group of the positive real numbers, the multiplicative group of the nonzero complex numbers, and the multiplicative group of the complex numbers of absolute value 1. In the construction of examples, it is well to bear in mind that *if* $\mathfrak{G}$ *is a Lie group and if* $\mathfrak{H}$ *is a subgroup of* $\mathfrak{G}$ *and also a submanifold of* $\mathfrak{G}$ *then* $\mathfrak{H}$ *is a Lie group.*

EXERCISES

1. Let X be a set with the structures of a real-analytic manifold M and a group G. Prove that the mapping μ_3: $M \times M \to M$ defined by $\mu_3(g_1, g_2) = g_1g_2^{-1}$ is real-analytic if and only if X is a Lie group. It follows that in the definition of a Lie group we may replace the two mappings μ_1 and μ_2 by the single mapping μ_3.

2. Let g_0 be a fixed element of a Lie group $\mathfrak{G}$. Prove that the mappings $g \to g_0 g$ and $g \to g g_0$ are real-analytic.

3. Prove that a submanifold of a real-analytic manifold is real-analytic.

4. Prove that the examples of Lie groups listed in this section are actually Lie groups.

5. Prove the last statement in this section.

7-2 THE GENERAL LINEAR GROUP

We shall give to the general linear group a real-analytic structure which will make it a Lie group. But first we shall discuss the topological structure of this group. We shall leave the proofs of many statements to the reader, including the proof of the following lemma.

Lemma 7-1. *Let U be an open set in V^n and C a compact subset of U. Then there is a neighborhood W of 0 such that if $v \in C$ and $w \in W$ then $v + w \in U$.*

The general linear group $GL(n)$ consists of the nonsingular linear transformations of V^n into V^n, and so it is a subset of the vector space $L(n)$ of *all* linear transformations of V^n into V^n. We may first topologize $L(n)$ and then give $GL(n)$ the topology induced on it by the topology in $L(n)$.

Let U be a neighborhood of 0 in V^n. Let C be a compact subset of V^n. Denote by $N(C, U)$ the set of all linear transformations in $L(n)$ which map C into U. If $\sigma_0 \in L(n)$, denote by $\sigma_0 + N(C, U)$ the set of all linear transformations $\sigma_0 + \sigma$ with $\sigma \in N(C, U)$. From Lemma 7-1 it follows that, if $\sigma_0 \in N(C, U)$, there is a neighborhood W of 0 in V^n such that $\sigma_0 + N(C, W) \subset N(C, U)$. The reader may now easily show that the family of sets $\{\sigma_0 + N(C, U)\}$ with $\sigma_0 \in L(n)$, C a compact subset of V^n, and U a neighborhood of 0 in V^n constitutes a basis for a topology in $L(n)$. This topology may then be used to define the topology on the subspace $GL(n)$.

The description of the topology on $L(n)$ which has just been given enjoys the advantage of not depending upon the choice of a basis for the vector space V^n. However, for numerical computations in $L(n)$ it is often necessary to represent its elements by matrices, and we shall now investigate how the topology in $L(n)$ appears in terms of a matrix representation of $L(n)$.

Let $\{v^1, \ldots, v^n\}$ be a basis of V^n. We then associate with $\sigma \in L(n)$ the matrix $\mathfrak{M}_\sigma = (m_{ij})$ for which

$$\sigma(v^j) = \sum_{i=1}^{n} m_{ij} v^i \qquad j = 1, \ldots, n$$

We shall see that a basis for the topology of $L(n)$ is provided by the sets of linear transformations σ whose matrices $\mathfrak{M}_\sigma$ belong to sets of the form $\{\mathfrak{M}_1 + \mathfrak{M}\}$, where $\mathfrak{M}_1$ is a fixed matrix and $\mathfrak{M} = (m_{ij})$ with

$$\sum_{i=1}^{n} \sum_{j=1}^{n} m_{ij}^2 < \epsilon^2$$

where $\epsilon > 0$. Actually it suffices to consider the case $\mathfrak{M}_1 = 0$ and to prove the following:

1. Given $N(C, U)$, there is an $\epsilon > 0$ such that whenever $\Sigma\Sigma m_{ij}^2 < \epsilon^2$ the linear transformation with matrix (m_{ij}) is in $N(C, U)$.
2. Given $\epsilon > 0$, there are sets C and U in V^n such that whenever $\sigma \in N(C, U)$ the matrix (m_{ij}) of σ satisfies $\Sigma\Sigma m_{ij}^2 < \epsilon^2$.

To prove point 1 we observe that the function

$$f(a_1v^1 + \cdots + a_nv^n) = \sqrt{a_1^2 + \cdots + a_n^2}$$

is bounded on the compact set C, say $f(v) \leqq K$ for $v \in C$. Also there is an $\eta > 0$ such that $f(v) < \eta$ implies $v \in U$. Take $\epsilon = \eta/K$.

To prove point 2, let U be the open subset of V^n consisting of the vectors v which satisfy $f(v) < \epsilon/\sqrt{n}$, and let C be the compact set of vectors v which satisfy $f(v) \leqq 1$. The details are left to the reader.

It is clear from our observations that the mapping

$$\sigma \to (m_{11}, \ldots, m_{1n}, m_{21}, \ldots, m_{nn})$$

where (m_{ij}) is the matrix of σ with respect to the basis $\{v^1, \ldots, v^n\}$, is a homeomorphism of $L(n)$ onto V^{n^2}. Consequently, it makes $L(n)$ into a coordinate neighborhood, and so it gives to $L(n)$ the structure of a differentiable or a real-analytic manifold. Since $GL(n)$ is an open subset of $L(n)$, $GL(n)$ may be given the structure of a submanifold of $L(n)$, and so it becomes a real-analytic manifold. Since a change of the basis $\{v^1, \ldots, v^n\}$ is effected by a nonsingular linear transformation of V^{n^2} onto itself, the choice of this basis is immaterial for the real-analytic structure of $GL(n)$. It is not difficult to show that the multiplicative group $GL(n)$ is a Lie group.

Theorem 7-1. *Let A be an affine algebraic variety in $L(n)$ (or E^{n^2}). If $A \cap GL(n)$ is a group, then $A \cap GL(n)$ is a Lie group.*

Proof. Let $\mathfrak{G} = A \cap GL(n)$. From Section 2-1, it suffices to prove that the jacobian matrix of the equations that define A has the same rank at all points of $\mathfrak{G}$, for then $\mathfrak{G}$ will be a differentiable variety in $L(n)$, and so it will be a submanifold of $GL(n)$.

If g_0 is any fixed element of $GL(n)$, the mapping $g \to g_0g$ is a real-analytic diffeomorphism of $GL(n)$. Hence if $\mathcal{J}(g)$ is the jacobian matrix at the point g of any finite set of functions on $GL(n)$, then

$$\mathfrak{J}(g_0 g) = \mathfrak{M}_0 \mathfrak{J}(g)$$

where $\mathfrak{M}_0$ is a nonsingular matrix which depends only on g_0. In particular, $\mathfrak{J}(g_0) = \mathfrak{M}_0 \mathfrak{J}(e)$, where e is the identity element of $GL(n)$, so that the rank of $\mathfrak{J}(g_0)$ is the same as the rank of $\mathfrak{J}(e)$. Since this is true for any element g_0 of $\mathfrak{G}$ [and even for any element of $GL(n)$], the theorem is proved.

EXERCISES

1. Prove the assertions made in this section.

2. Prove that the orthogonal group $O(n)$ is a Lie group.

3. Prove that the group of all $n \times n$ matrices whose determinants are equal to 1 is a Lie group.

4. Prove that the set of all $n \times n$ matrices of the form

$$\begin{pmatrix} 1 & a_{12} & \cdots & a_{1n} \\ 0 & 1 & \cdots & a_{2n} \\ \cdots & \cdots & \cdots & \cdots \\ 0 & 0 & \cdots & 1 \end{pmatrix}$$

is a Lie group.

5. Prove that the set of all $n \times n$ matrices of the form

$$\begin{pmatrix} a_{11} & a_{12} & \cdots & a_{1n} \\ 0 & a_{22} & \cdots & a_{2n} \\ \cdots & \cdots & \cdots & \cdots \\ 0 & 0 & \cdots & a_{nn} \end{pmatrix}$$

is a Lie group.

7-3 ONE-PARAMETER SUBGROUPS IN $GL(n)$

The general linear group serves as a useful example for illustrating the features of Lie groups. Later we shall see how the theory may be generalized for arbitrary Lie groups.

Let $\mathfrak{G}$ and $\mathfrak{H}$ be two Lie groups. A mapping $\sigma: \mathfrak{G} \to \mathfrak{H}$ is called a homomorphism of $\mathfrak{G}$ into $\mathfrak{H}$ if the following conditions prevail:

1. σ is a group homomorphism.
2. σ is a regular real-analytic mapping.

If $\sigma: \mathfrak{G} \to \mathfrak{H}$ is a homomorphism, then $\sigma(\mathfrak{G})$ is a subgroup of $\mathfrak{H}$. Although $\sigma(\mathfrak{G})$ is also a submanifold of $\mathfrak{H}$ and therefore a Lie group, we shall not need this fact now.

In particular, we may take for $\mathfrak{G}$ the real line V^1 under addition. The image $\sigma(V^1)$ under a homomorphism $\sigma: V^1 \to \mathfrak{H}$ will be called a *one-parameter subgroup* of $\mathfrak{H}$. We have the rule $\sigma(t_1 + t_2) = \sigma(t_1)\sigma(t_2)$.

If σ_* is the linear transformation induced by σ on the tangent space, the regularity of σ is equivalent to the statement that $\sigma_*(\partial/\partial t)$ is not zero.

In dealing with $L(n)$ we shall use a matrix representation and regard $L(n)$ as the ring of all $n \times n$ matrices. $GL(n)$ is then identified with the multiplicative group of all nonsingular matrices in $L(n)$. We shall describe a method of constructing one-parameter subgroups of $GL(n)$. The method is a rather formal procedure employing infinite series of matrices. Later we shall analyze its abstract features in order to see how it can be made to work in general Lie groups.

Let $\mathfrak{A} \in L(n)$, and consider the infinite series

$$\mathcal{E} + \mathfrak{A} + \frac{1}{2!}\mathfrak{A}^2 + \frac{1}{3!}\mathfrak{A}^3 + \cdots$$

where $\mathcal{E}$ is the identity matrix. This infinite series is to be interpreted as a sequence of matrices in $L(n)$, the terms of the sequence being the partial sums of the series. We shall prove that this infinite series converges in the topology of $L(n)$ to a unique limit matrix which we denote by $\exp \mathfrak{A}$. This is equivalent to asserting that each component of the partial-sum matrix converges as a sequence of real numbers.

Let $\mathcal{E} = (\delta_{ij})$, $\mathfrak{A} = (a_{ij})$, and $\mathfrak{A}^m = (a_{ij}{}^{(m)})$ for $m = 2, 3, \ldots$. Then the component of $\exp \mathfrak{A}$ in the ith row and jth column, if it exists, is the limit of the infinite series

$$\delta_{ij} + a_{ij} + \frac{1}{2!}a_{ij}{}^{(2)} + \frac{1}{3!}a_{ij}{}^{(3)} + \cdots$$

We must show that this series converges. Let $a = \max\{|a_{ij}|\}$ and $a^{(m)} = \max\{|a_{ij}{}^{(m)}|\}$ for $m = 2, 3, \ldots$. Since

$$a_{ij}{}^{(m)} = \sum_{k=1}^{n} a_{ik}{}^{(m-1)}a_{kj}$$

it follows that

$$a^{(m)} \leqq naa^{(m-1)}$$

Using this repeatedly, we obtain

$$a^{(m)} \leqq (na)^m$$

Hence

$$\frac{1}{m!}|a_{ij}{}^{(m)}| \leqq \frac{(na)^m}{m!}$$

It follows that the series for the components of $\exp \mathfrak{A}$ converge absolutely since they may be compared with the convergent series

$$1 + na + \frac{1}{2!}(na)^2 + \frac{1}{3!}(na)^3 + \cdots$$

Theorem 7-2. *$\mathfrak{A} \to \exp \mathfrak{A}$ is a real-analytic mapping of $L(n)$ into $L(n)$ which is regular at 0 and maps a neighborhood of 0 homeomorphically onto a neighborhood of $\mathcal{E}$.*

Proof. We shall use the notation of the previous paragraph. We observe that $a_{ij}^{(m)}$ is a polynomial in the variables $a_{11}, a_{12}, \ldots, a_{nn}$. Hence the partial sums of the series

$$\delta_{ij} + a_{ij} + \frac{1}{2!} a_{ij}^{(2)} + \frac{1}{3!} a_{kj}^{(3)} + \cdots$$

are also polynomials in $a_{11}, a_{12}, \ldots, a_{nn}$. It follows that the components of $\exp \mathfrak{A}$ are convergent power series in the variables $a_{11}, a_{12}, \ldots, a_{nn}$ and hence that $\mathfrak{A} \to \exp \mathfrak{A}$ is a real-analytic mapping.

Since the points of $L(n)$ have the coordinates

$$\begin{pmatrix} x_{11} & \cdots & x_{1n} \\ \cdot & \cdots & \cdot \\ x_{n1} & \cdots & x_{nn} \end{pmatrix}$$

the vectors

$$\left(\frac{\partial}{\partial x_{11}}\right)_0, \ldots, \left(\frac{\partial}{\partial x_{nn}}\right)_0$$

are a basis of the tangent space to $L(n)$ at 0. We shall show that, if $\exp_*$ is the linear transformation induced by exp on this tangent space, then

$$\exp_* \left(\frac{\partial}{\partial x_{11}}\right)_0, \ldots, \exp_* \left(\frac{\partial}{\partial x_{nn}}\right)_0$$

are linearly independent. It follows that $\exp_*$ is an isomorphism of the tangent space to $L(n)$ at 0 onto the tangent space to $L(n)$ at $\mathcal{E}$. We have seen in Section 2-7 that such a mapping provides us with a local homeomorphism.

$(\partial/\partial x_{kl})_0$ is the tangent vector to the curve $\gamma(t) = t\mathfrak{A}_{kl}$, where $\mathfrak{A}_{kl}$ is the matrix with 0 everywhere except for a 1 in the kth row and lth column. $\exp \gamma(t)$ is easily computed, and we obtain a matrix which is the same as the identity matrix $\mathcal{E}$ except that in the kth row and lth column there is

$$\begin{cases} e^t & \text{if } k = l \\ t & \text{if } k \neq l \end{cases}$$

Hence

$$\exp_* \left(\frac{\partial}{\partial x_{kl}}\right)_0 = \left(\frac{\partial}{\partial x_{kl}}\right)_{\mathcal{E}}$$

for all k and l. This completes the proof of the theorem.

Our aim is to show that the matrix $\exp \mathcal{A}$ is nonsingular, and consequently the mapping $t \to \exp t\mathcal{A}$ may be used to define a one-parameter subgroup of the general linear group.

Lemma 7-2. *If $\mathcal{A}$ and $\mathcal{B}$ are matrices from $L(n)$ for which $\mathcal{A}\mathcal{B} = \mathcal{B}\mathcal{A}$, then*

$$\exp(\mathcal{A} + \mathcal{B}) = \exp \mathcal{A} \exp \mathcal{B}$$

Proof. Denote by $\mathcal{R}_m$ the expression

$$\sum_{k=0}^{2m} \frac{1}{k!} (\mathcal{A} + \mathcal{B})^k - \left(\sum_{k=0}^{m} \frac{1}{k!} \mathcal{A}^k\right)\left(\sum_{k=0}^{m} \frac{1}{k!} \mathcal{B}^k\right).$$

The lemma will be proved if we can prove that the sequence $\mathcal{R}_1$, $\mathcal{R}_2$, . . . converges to 0. The binomial theorem is valid for $(\mathcal{A} + \mathcal{B})^k$. Hence

$$\begin{aligned}\mathcal{R}_m &= \sum_{j=0}^{2m} \sum_{k=0}^{j} \frac{1}{k!(j-k)!} \mathcal{A}^k\mathcal{B}^{j-k} - \sum_{k=0}^{m} \sum_{j=0}^{m} \frac{1}{k!j!} \mathcal{A}^k\mathcal{B}^j \\ &= \sum_{\substack{k+j \leq 2m \\ k>m \text{ or } j>m}} \frac{1}{k!} \mathcal{A}^k \frac{1}{j!} \mathcal{B}^j\end{aligned}$$

Denote by c the maximum of the absolute values of the components of the matrices $\mathcal{A}$ and $\mathcal{B}$. We have seen that the components of the matrices $\mathcal{A}^k$ and $\mathcal{B}^k$ are then bounded by $(nc)^k$. Hence the components of $(1/k!j!)\, \mathcal{A}^k\mathcal{B}^j$ are bounded by $n(nc)^{k+j}/k!j!$. There are fewer than m^2 such terms in the expression for $\mathcal{R}_m$, so that each component of $\mathcal{R}_m$ is bounded by $m^2n(nc)^{2m}/m!$. This bound may be made arbitrarily small by choosing m sufficiently large, and so the lemma is proved.

Theorem 7-3. $\exp \mathcal{A}$ *is nonsingular.*

Proof. Let t be a real number and $f(t) = \det \exp t\mathcal{A}$. Then $f(t)$ is a continuous function on E^1. Since $f(0) = 1$, there is an $\epsilon > 0$ such that $|f(t)| > \frac{1}{2}$ for $|t| < 2\epsilon$. Thus for $|t| < 2\epsilon$ we know that the matrix $\exp t\mathcal{A}$ is nonsingular.

Suppose that for some real number u the matrix $\exp u\mathcal{A}$ is nonsingular. Let $|t - u| < 2\epsilon$. Since

$$\exp t\mathcal{A} = \exp[(t-u)\mathcal{A}] \exp u\mathcal{A}$$

is the product of nonsingular matrices, $\exp t\mathcal{A}$ is nonsingular. Thus we may conclude that the matrices

$$\exp \epsilon\mathcal{A}, \exp 2\epsilon\mathcal{A}, \exp 3\epsilon\mathcal{A}, \ldots$$

are all nonsingular and also the matrices

$$\exp(-\epsilon)\mathfrak{A},\ \exp(-2\epsilon)\mathfrak{A},\ \exp(-3\epsilon)\mathfrak{A},\ \ldots$$

It follows that $\exp t\mathfrak{A}$ is nonsingular for all values of t. In particular, $\exp \mathfrak{A}$ is nonsingular.

Since $t \to \exp t\mathfrak{A}$ is a group homomorphism of V^1 into $GL(n)$ and also a real-analytic mapping, it provides us with a one-parameter subgroup of $GL(n)$. There is such a one-parameter subgroup for each matrix $\mathfrak{A}$ in $L(n)$. We shall see later that there are no other one-parameter subgroups of $GL(n)$.

EXERCISES

1. If $\mathfrak{A}$ is the diagonal matrix

$$\mathfrak{A} = \begin{pmatrix} a_1 & 0 & \cdots & 0 \\ 0 & a_2 & \cdots & 0 \\ \cdots & \cdots & \cdots & \cdots \\ 0 & 0 & \cdots & a_n \end{pmatrix}$$

compute $\exp t\mathfrak{A}$.

2. Show that if $\mathfrak{A}$ is a matrix of the form

$$\mathfrak{A} = \begin{pmatrix} 0 & a_{12} & a_{13} & \cdots & a_{1n} \\ 0 & 0 & a_{23} & \cdots & a_{2n} \\ \cdots & \cdots & \cdots & \cdots & \cdots \\ 0 & 0 & 0 & \cdots & 0 \end{pmatrix}$$

then $\exp t\mathfrak{A}$ is a polynomial in t.

3. Prove that, if $\mathcal{P}$ is a nonsingular matrix, then

$$\mathcal{P}(\exp \mathfrak{A})\mathcal{P}^{-1} = \exp(\mathcal{P}\mathfrak{A}\mathcal{P}^{-1})$$

4.[1] Prove that, if v is an eigenvector of $\mathfrak{A}$ with the eigenvalue c, then v is an eigenvector of $\exp \mathfrak{A}$ with the eigenvalue e^c.

5.[1] Prove that

$$\det \exp \mathfrak{A} = \exp(\text{trace } \mathfrak{A})$$

7-4 LEFT-INVARIANT VECTOR FIELDS

Let $\mathfrak{G}$ be a Lie group. Let $t \to \gamma(t)$ be a homomorphism of the Lie group V^1 into the Lie group $\mathfrak{G}$. This mapping defines the one-parameter subgroup of $\mathfrak{G}$ which consists of the elements $\{\gamma(t)\}$, and at the same time it describes a differentiable parametrized curve γ in $\mathfrak{G}$, in the extended sense introduced in Section 4-4.

If g is any fixed element of $\mathfrak{G}$, the mapping $t \to g\gamma(t)$ is real-analytic

[1] For those readers who are acquainted with sufficient linear algebra.

and therefore defines a differentiable parametrized curve in $\mathfrak{G}$ (but not a one-parameter subgroup) which we denote by $g\gamma$. If we allow g to range over the elements of $\mathfrak{G}$, we obtain a family $\{g\gamma\}$ of curves which completely covers $\mathfrak{G}$. Indeed, if h is a point of $\mathfrak{G}$, many different curves of this family pass through h. We shall now prove that all the curves of the family $\{g\gamma\}$ which pass through h have the same tangent vector at h.

Theorem 7-4. *If $t \to \gamma(t)$ is a homomorphism of V^1 into a Lie group $\mathfrak{G}$ and if $h = g_1\gamma(t_1) = g_2\gamma(t_2)$, where $g_1, g_2 \in \mathfrak{G}$, then the curves $g_1\gamma$ and $g_2\gamma$ have the same tangent vector at h.*

Proof. Let us first assume that $h = g\gamma(t_1) = \gamma(t_2)$. Let U be a coordinate neighborhood of h with the coordinates $(x_1, \ldots, x_n)$. On U we may write

$$\gamma(t) = (x_1(t), \ldots, x_n(t))$$

with t in a neighborhood of t_2. But we also have

$$\gamma(t + t_2 - t_1) = \gamma(t_2)\gamma(t_1)^{-1}\gamma(t) = g\gamma(t)$$

and so we may write

$$g\gamma(t) = (x_1(t + t_2 - t_1), \ldots, x_n(t + t_2 - t_1))$$

with t in a neighborhood of t_1. From these expressions it is apparent that γ and $g\gamma$ have the same tangent vector at h.

More generally, if $h = g_1\gamma(t_1) = g_2\gamma(t_2)$, then

$$g_2^{-1}h = g_2^{-1}g_1\gamma(t_1) = \gamma(t_2)$$

Consequently $g_2^{-1}g_1\gamma$ and γ have the same tangent vector at $g_2^{-1}h$. It follows that $g_1\gamma$ and $g_2\gamma$ have the same tangent vector at h.

Let $\gamma\colon V^1 \to \mathfrak{G}$ be a one-parameter subgroup of a Lie group $\mathfrak{G}$. If $h = g\gamma(t)$, define $\chi(h)$ = tangent vector to $g\gamma$ at h. According to Theorem 7-4, the vector field χ is well-defined on $\mathfrak{G}$. For each $g \in \mathfrak{G}$, denote by σ^g the mapping $\sigma^g\colon \mathfrak{G} \to \mathfrak{G}$ given by $\sigma^g(h) = gh$. Let $\sigma^g{}_*$ denote the linear transformation of the tangent space induced by σ^g. Then the vector field χ satisfies the relation $\sigma^g{}_*\circ\chi = \chi\circ\sigma^g$. Any vector field on $\mathfrak{G}$ which satisfies this relation for all $g \in \mathfrak{G}$ will be said to be *left-invariant.* We have thus found that each one-parameter subgroup γ of the Lie group $\mathfrak{G}$ determines a left-invariant vector field χ on $\mathfrak{G}$. The flow of this vector field χ consists of the integral manifolds $g\gamma$.

Conversely, let χ be a left-invariant vector field on $\mathfrak{G}$. It is reasonable to conjecture that the flow determined by χ must arise by left multiplication from some one-parameter subgroup of $\mathfrak{G}$. The difficulty that we face if we attempt to prove this conjecture is that we only know the existence of integral manifolds for χ in the neighborhood of a point. Thus we have only a mapping $t \to \gamma(t)$ of an interval of V^1 into $\mathfrak{G}$, whereas we need a mapping of all of V^1 into $\mathfrak{G}$. Nevertheless, we shall see that the group structure on $\mathfrak{G}$ enables us to extend the mapping γ to all of V^1.

Let χ be a real-analytic left-invariant vector field on $\mathfrak{G}$. Let γ be an integral manifold for χ which passes through the identity element of $\mathfrak{G}$. We may assume that $\gamma(0)$ is the identity element of $\mathfrak{G}$. Then there is a positive number ϵ such that $\gamma(t)$ is defined for $|t| < \epsilon$. Let $|t_0| < \epsilon$ and $g = \gamma(t_0)$. Let σ denote left multiplication by g, and let σ_* be the linear transformation that σ induces on the tangent space at a point. Then the tangent vector to the curve $g\gamma$ at $g\gamma(t)$ is $(\sigma_*\circ\chi)[\gamma(t)]$. Since χ is left-invariant, $(\sigma_*\circ\chi)[\gamma(t)] = \chi[\sigma\circ\gamma(t)] = \chi[g\gamma(t)]$. Hence $g\gamma$ is also an integral manifold for χ. Both γ and $g\gamma$ pass through the point $g = \gamma(t_0) = g\gamma(0)$. The uniqueness statement of the theorem on integral manifolds tells us that $\gamma(t_0 + t) = g\gamma(t)$ for those values of t for which $|t_0 + t| < \epsilon$ and $|t| < \epsilon$. Thus we have proved that for $|t_0| < \epsilon$, $|t| < \epsilon$, and $|t_0 + t| < \epsilon$

$$\gamma(t_0 + t) = \gamma(t_0)\gamma(t)$$

We shall now show how to extend γ so that it is defined for $|t| < 2\epsilon$. Let $\epsilon \leqq t < 2\epsilon$. Then we may write $t = t_1 + t_2$, with $t_1 < \epsilon$ and $t_2 < \epsilon$. We wish to define $\gamma(t) = \gamma(t_1)\gamma(t_2)$. Suppose that $t = t_1' + t_2'$, with $t_1' < \epsilon$ and $t_2' < \epsilon$. Then $t_1 - t_1' = t_2' - t_2$ and $|t_1 - t_1'| < \epsilon$. Hence $\gamma(t_1)\gamma(-t_1') = \gamma(t_2')\gamma(-t_2)$ or $\gamma(t_1')\gamma(t_2') = \gamma(t_1)\gamma(t_2)$. Thus our definition for $\epsilon \leqq t < 2\epsilon$ is legitimate. We may similarly define $\gamma(t) = \gamma(t_1)\gamma(t_2)$ for $-2\epsilon < t \leqq -\epsilon$ if $t = t_1 + t_2$, with $t_1 > -\epsilon$ and $t_2 > -\epsilon$. The left invariance of χ assures us that the extended curve γ will still be an integral manifold for χ.

The construction by which γ was extended from the interval $(-\epsilon, \epsilon)$ to the interval $(-2\epsilon, 2\epsilon)$ may be repeated in order to produce an integral manifold γ defined for all values of t. Then $t \to \gamma(t)$ will be a group homomorphism of V^1 into $\mathfrak{G}$. It is not difficult to prove that, since χ is a real-analytic vector field [and consequently in any suitable coordinate neighborhood the coefficients of the system of differential equations of which $\gamma(t)$ is a solution are real-analytic functions], the mapping $t \to \gamma(t)$ is real-analytic. Hence this mapping defines a one-parameter subgroup of $\mathfrak{G}$, and χ is the left-invariant vector field which is determined by this one-parameter subgroup. Thus we have proved the following result.

Theorem 7-5. *A homomorphism $t \to \gamma(t)$ of the Lie group V^1 into a Lie group $\mathfrak{G}$ determines, through left multiplication by elements of $\mathfrak{G}$, a left-invariant vector field on $\mathfrak{G}$. Conversely, every left-invariant vector field on $\mathfrak{G}$ is determined by a homomorphism of V^1 into $\mathfrak{G}$.*

It is easy to see that a left-invariant vector field on $\mathfrak{G}$ is completely determined by its value at the identity element of $\mathfrak{G}$. The preceding theorem then says that a homomorphism $t \to \gamma(t)$ of V^1 into $\mathfrak{G}$ is completely determined by the tangent vector to the curve $\gamma(t)$ at $t = 0$. Since any vector from the tangent space to $\mathfrak{G}$ at the identity element may be used to generate a left-invariant vector field on $\mathfrak{G}$, it follows that such vector fields are in one-to-one correspondence with the vectors of the tangent space to $\mathfrak{G}$ at the identity element. In the next section we shall see what this means for the general linear group.

EXERCISES

1. Prove that left-invariant vector field χ on $\mathfrak{G}$ is completely determined by its value at the identity element of $\mathfrak{G}$.

2. Prove that, if χ is a left-invariant vector field, χ is real-analytic; that is, as a mapping of $\mathfrak{G}$ into the tangent bundle $\mathfrak{T}(\mathfrak{G})$, it is real-analytic.

7-5 THE EXPONENTIAL MAPPING IN $GL(n)$

We shall determine the left-invariant vector fields on $GL(n)$. A point

$$\mathfrak{X} = \begin{pmatrix} x_{11} & \cdots & x_{1n} \\ \cdot & \cdots & \cdot \\ x_{n1} & \cdots & x_{nn} \end{pmatrix}$$

of $GL(n)$ has the coordinates $x_{11}, x_{12}, \ldots, x_{nn}$. Consequently

$$\left\{\left(\frac{\partial}{\partial x_{11}}\right)_{\mathfrak{X}}, \left(\frac{\partial}{\partial x_{12}}\right)_{\mathfrak{X}}, \ldots, \left(\frac{\partial}{\partial x_{nn}}\right)_{\mathfrak{X}}\right\}$$

is a basis of the tangent space to $GL(n)$ at the point $\mathfrak{X}$. A vector in the tangent space to $GL(n)$ at $\mathfrak{X}$ then has the unique representation

$$\alpha = a_{11}\left(\frac{\partial}{\partial x_{11}}\right)_{\mathfrak{X}} + \cdots + a_{nn}\left(\frac{\partial}{\partial x_{nn}}\right)_{\mathfrak{X}}$$

It is convenient to arrange the coefficients of α in the form of a matrix

$$\begin{pmatrix} a_{11} & \cdots & a_{1n} \\ \cdot & \cdots & \cdot \\ a_{n1} & \cdots & a_{nn} \end{pmatrix}$$

and we see that the tangent space to $GL(n)$ at a point has the algebraic structure of $L(n)$, the space of all $n \times n$ matrices. We shall therefore speak of the tangent vector $\mathcal{A} = (a_{ij})$ to $GL(n)$ at the point $\mathcal{X}$ instead of the tangent vector α.

Theorem 7-6. *Let $\mathcal{A}$ be a tangent vector to $GL(n)$ at the identity element $\mathcal{E}$. Let χ be the left-invariant vector field on $GL(n)$ for which $\chi(\mathcal{E}) = \mathcal{A}$. Then at each point $\mathcal{X}$ of $GL(n)$*

$$\chi(\mathcal{X}) = \mathcal{X}\mathcal{A}$$

Proof. We have seen in the previous section that there is a unique one-parameter subgroup $\{\gamma(t)\}$ such that the tangent vector to γ at $\mathcal{E}$ is $\mathcal{A}$. Then the tangent vector to $\mathcal{X}\gamma$ at $\mathcal{X}$ is $\mathcal{X}\mathcal{A}$. But this is the vector $\chi(\mathcal{X})$.

Theorem 7-7. *Let $\mathcal{A}$ be a tangent vector to $GL(n)$ at the identity element $\mathcal{E}$. Let $\{\gamma(t)\}$ be the one-parameter subgroup of $GL(n)$ which has the tangent vector $\mathcal{A}$ at $\mathcal{E}$. Then*

$$\gamma(t) = \exp t\mathcal{A}$$

Proof. It suffices to demonstrate that, if $\gamma_1(t) = \exp t\mathcal{A}$, the tangent vector to γ_1 at $\mathcal{E}$ is $\mathcal{A}$. For we know that $\{\gamma_1(t)\}$ is a one-parameter subgroup of $GL(n)$.

Let $\exp_*$ be the linear transformation which the mapping exp induces on the tangent space to $L(n)$ at 0. We have proved that

$$\exp_* \left(\frac{\partial}{\partial x_{ij}}\right)_0 = \left(\frac{\partial}{\partial x_{ij}}\right)_{\mathcal{E}}$$

for all i and j. Let $\Gamma(t) = t\mathcal{A}$. Then Γ is a differentiable parametrized curve in $L(n)$ whose tangent vector at $t = 0$ is

$$\alpha = \sum_{i,j=1}^{n} a_{ij} \left(\frac{\partial}{\partial x_{ij}}\right)_0$$

where $\mathcal{A} = (a_{ij})$. Hence

$$\exp_* \alpha = \sum_{i,j=1}^{n} a_{ij} \left(\frac{\partial}{\partial x_{ij}}\right)_{\mathcal{E}}$$

is the tangent vector to the curve $\exp \Gamma$ at $\mathcal{E}$, which was to be proved.

We see that we may regard the exponential mapping $\mathcal{A} \to \exp \mathcal{A}$ as a mapping of the tangent space to $GL(n)$ at $\mathcal{E}$ into $GL(n)$. If we adopt this viewpoint, we shall be able to generalize the exponential mapping to any Lie group.

7-6 THE EXPONENTIAL MAPPING IN LIE GROUPS

It is now clear how to define the analogue for Lie groups of the mapping $\mathfrak{A} \to \exp \mathfrak{A}$ for $GL(n)$. Let $\mathfrak{G}$ be a Lie group, and let T be the tangent space to $\mathfrak{G}$ at the identity element. Let U be a coordinate neighborhood of the identity element of $\mathfrak{G}$, and let $(x_1, \ldots, x_n)$ be the coordinates of the points of U. Then an element α of T has the form

$$\alpha = a_1\left(\frac{\partial}{\partial x_1}\right)_e + \cdots + a_n\left(\frac{\partial}{\partial x_n}\right)_e$$

where e is the identity element of $\mathfrak{G}$. If we take $(a_1, \ldots, a_n)$ to be the coordinates of α, then T becomes a differentiable manifold. The structure of T as a differentiable manifold does not depend upon which coordinate neighborhood U we use, and it makes T into a Lie group.

The exponential mapping for $\mathfrak{G}$ will then be defined as a mapping of T into $\mathfrak{G}$ as follows: Let α be an element of T. We have seen that there is a unique one-parameter subgroup $\gamma(t)$ such that the curve γ has the tangent vector α at $t = 0$. Define

$$\exp \alpha = \gamma(1)$$

It is possible to prove that the mapping $\alpha \to \exp \alpha$ enjoys the properties which we have observed in the special case of the general linear group. Specifically, they are as follows:

1. It is a real-analytic mapping of T into $\mathfrak{G}$.
2. It maps a neighborhood of the zero element of T in a one-to-one manner onto a neighborhood of the identity element of $\mathfrak{G}$.

The proofs of these properties are too involved to be given here.

EXERCISE

Prove that, if α is in T and t_1 and t_2 are real numbers, then

$$\exp (t_1 + t_2)\alpha = (\exp t_1\alpha)(\exp t_2\alpha)$$

7-7 THE GLOBAL BEHAVIOR OF ONE-PARAMETER SUBGROUPS

In this section we shall examine a number of examples of Lie groups which illustrate special features of the topics introduced in this chapter.

Example 1. The 2-torus T^2 is the factor group V^2/D, where D is the subgroup consisting of those vectors of V^2 which have integral components. It is not difficult to verify that T^2 is a Lie group and that the natural group homomorphism $\sigma: V_2 \to T^2$ is a homomorphism of the Lie group V^2 onto the Lie group T^2.

V^2 may be regarded as the tangent space to T^2 at the identity element. If $v \in V^2$, then $\gamma(t) = \sigma(tv)$ is the one-parameter subgroup of T^2 with the tangent vector v at the identity element. Hence $\exp v = \sigma(v)$ in this case. It is therefore apparent that the exponential mapping in T^2 is not globally one-to-one.

It may be objected that T^2 is not a simply connected space and that perhaps the exponential mapping must be globally one-to-one in simply connected spaces. The next example shows that this is not the case.

Example 2. Let the set $\mathfrak{S}$ consist of all 4×4 matrices of the form

$$\mathcal{S}(u, v, w) = \begin{pmatrix} \cos w & \sin w & 0 & u \\ -\sin w & \cos w & 0 & v \\ 0 & 0 & 1 & w \\ 0 & 0 & 0 & 1 \end{pmatrix}$$

where u, v, w are real numbers. Then

$$\mathcal{S}(u_1, v_1, w_1)\mathcal{S}(u_2, v_2, w_2) = \mathcal{S}(u, v, w_1 + w_2)$$

where

$$\begin{aligned} u &= u_2 \cos w_1 + v_2 \sin w_1 + u_1 \\ v &= -u_2 \sin w_1 + v_2 \cos w_1 + v_1 \end{aligned}$$

Hence $\mathfrak{S}$ is a subgroup of $GL(4)$. The mapping σ defined by $\sigma(u, v, w) = \mathcal{S}(u, v, w)$ is a differentiable one-to-one mapping of E^3 into $GL(4)$. Let σ_* be the induced linear transformation of the tangent space to E^3. Then

$$\sigma_*\left(\frac{\partial}{\partial u}\right) = \frac{\partial}{\partial x_{14}}$$

$$\sigma_*\left(\frac{\partial}{\partial v}\right) = \frac{\partial}{\partial x_{24}}$$

$$\sigma_*\left(\frac{\partial}{\partial w}\right) = -\sin w \frac{\partial}{\partial x_{11}} + \cos w \frac{\partial}{\partial x_{12}} - \cos w \frac{\partial}{\partial x_{21}} - \sin w \frac{\partial}{\partial x_{22}} + \frac{\partial}{\partial x_{34}}$$

Hence $\sigma_*(\partial/\partial u)$, $\sigma_*(\partial/\partial v)$, $\sigma_*(\partial/\partial w)$ are linearly independent. Thus σ is regular, and we have proved that $\mathfrak{S}$ is a 3-dimensional submanifold of $GL(4)$. Hence $\mathfrak{S}$ is a Lie group.

It follows from our computations that the tangent space to $\mathfrak{S}$ at the identity element has the basis $\{\alpha_1, \alpha_2, \alpha_3\}$, where

$$\alpha_1 = \left(\frac{\partial}{\partial x_{14}}\right)_{\mathcal{E}}$$

$$\alpha_2 = \left(\frac{\partial}{\partial x_{24}}\right)_{\mathcal{E}}$$

$$\alpha_3 = \left(\frac{\partial}{\partial x_{12}}\right)_{\mathcal{E}} - \left(\frac{\partial}{\partial x_{21}}\right)_{\mathcal{E}} + \left(\frac{\partial}{\partial x_{34}}\right)_{\mathcal{E}}$$

Every one-parameter subgroup of $\mathfrak{S}$ is also a one-parameter subgroup of $GL(4)$. Consequently we may employ the conventions and results established in Section 7-5.

If $\gamma(t)$ is the one-parameter subgroup of $\mathfrak{S}$ whose tangent vector at the identity element is $a\alpha_1 + b\alpha_2 + c\alpha_3$, then $\gamma(t) = \exp t\mathfrak{A}$, where

$$\mathfrak{A} = \begin{pmatrix} 0 & c & 0 & a \\ -c & 0 & 0 & b \\ 0 & 0 & 0 & c \\ 0 & 0 & 0 & 0 \end{pmatrix}$$

It is not difficult to show by direct computation that

$$\gamma(t) = \begin{pmatrix} \cos ct & \sin ct & 0 & u(t) \\ -\sin ct & \cos ct & 0 & v(t) \\ 0 & 0 & 1 & ct \\ 0 & 0 & 0 & 1 \end{pmatrix}$$

where $u(t)$ and $v(t)$ are real-analytic functions of t.

Suppose that c is not zero. Since $\gamma(2\pi/c) = \gamma(\pi/c)\gamma(\pi/c)$,

$$\begin{pmatrix} 1 & 0 & 0 & u(2\pi/c) \\ 0 & 1 & 0 & v(2\pi/c) \\ 0 & 0 & 1 & 2\pi \\ 0 & 0 & 0 & 1 \end{pmatrix} = \begin{pmatrix} -1 & 0 & 0 & u(\pi/c) \\ 0 & -1 & 0 & v(\pi/c) \\ 0 & 0 & 1 & \pi \\ 0 & 0 & 0 & 1 \end{pmatrix}^2$$

It follows that $u(2\pi/c) = v(2\pi/c) = 0$. We see therefore that every one-parameter subgroup of $\mathfrak{S}$ whose tangent vector $a\alpha_1 + b\alpha_2 + c\alpha_3$ at the identity element has $c \neq 0$ passes through the point

$$\begin{pmatrix} 1 & 0 & 0 & 0 \\ 0 & 1 & 0 & 0 \\ 0 & 0 & 1 & 2\pi \\ 0 & 0 & 0 & 1 \end{pmatrix}$$

in $\mathfrak{S}$. Consequently the exponential mapping in $\mathfrak{S}$ is not one-to-one.

Example 3. Let the set $\mathfrak{S}$ consist of all 3×3 matrices of the form

$$\begin{pmatrix} 1 & u & v \\ 0 & 1 & w \\ 0 & 0 & 1 \end{pmatrix}$$

We may show by the methods used in Example 2 that $\mathfrak{S}$ is a Lie group, that the tangent space to $\mathfrak{S}$ at the identity element is spanned by the vectors $(\partial/\partial x_{12})_{\mathcal{E}}$, $(\partial/\partial x_{13})_{\mathcal{E}}$, $(\partial/\partial x_{23})_{\mathcal{E}}$, and that the one-parameter subgroups of $\mathfrak{S}$ are given by $\gamma(t) = \exp t\mathfrak{A}$, where $\mathfrak{A}$ has the form

$$\mathfrak{A} = \begin{pmatrix} 0 & a & b \\ 0 & 0 & c \\ 0 & 0 & 0 \end{pmatrix}$$

Thus by direct computation

$$\gamma(t) = \begin{pmatrix} 1 & at & bt + \frac{1}{2}act^2 \\ 0 & 1 & ct \\ 0 & 0 & 1 \end{pmatrix}$$

Since for any numbers u, v, w there is one and only one set of numbers a, b, c such that

$$u = a \qquad v = b + \tfrac{1}{2}ac \qquad w = c$$

it follows that the exponential mapping in $\mathfrak{S}$ maps the tangent space at the identity element in a one-to-one manner globally onto $\mathfrak{S}$.

Example 4. Let $O^+(n)$ be the subgroup of the orthogonal matrices which have their determinants equal to $+1$. Since $O^+(n)$ is a subgroup of $GL(n)$ and is an algebraic variety, it is a Lie group. We wish to determine the tangent space to $O^+(n)$ at the identity element. Let

$$\gamma(t) = \begin{pmatrix} x_{11}(t) & \cdots & x_{1n}(t) \\ \cdot & \cdots & \cdot \\ x_{n1}(t) & \cdots & x_{nn}(t) \end{pmatrix}$$

be any differentiable parametrized curve in $O^+(n)$ with

$$\gamma(0) = \mathcal{E} = (\delta_{ij})$$

Then we know that

$$\sum_{j=1}^{n} x_{ij}(t)x_{kj}(t) = \delta_{ik}$$

for $i, k = 1, \ldots, n$. If we differentiate this relation with respect to t, set $t = 0$, and remember that $x_{ij}(0) = \delta_{ij}$, we obtain

$$x'_{ik}(0) + x'_{ki}(0) = 0$$

for $i, k = 1, \ldots, n$. Hence the tangent vector $(x'_{ik}(0))$ to γ at $t = 0$ is skew-symmetric.

The vector space of all $n \times n$ skew-symmetric matrices has dimension $\frac{1}{2}n(n-1)$. We shall produce a set of $\frac{1}{2}n(n-1)$ tangent vectors to $O^+(n)$ which constitute a basis of this vector space. Thus we shall prove that the vector space of all $n \times n$ skew-symmetric matrices is the tangent space to $O^+(n)$ at the identity element.

Consider the curve

$$\begin{array}{rl} & \begin{array}{ccccccccccc} & & & i & & & & j & & & \end{array} \\ \gamma_{ij}(t) = & \left(\begin{array}{ccccccccccc} \cdot & & & & & & & & & & \\ & \cdot & & & & & & & & & \\ & & \cdot & & & & & & & & \\ & & & \cos t & & & & \sin t & & & \\ & & & & \cdot & & & & & & \\ & & & & & \cdot & & & & & \\ & & & & & & \cdot & & & & \\ & & & -\sin t & & & & \cos t & & & \\ & & & & & & & & \cdot & & \\ & & & & & & & & & \cdot & \\ & & & & & & & & & & \cdot \end{array}\right) \begin{array}{c} \\ \\ \\ i \\ \\ \\ \\ j \\ \\ \\ \\ \end{array} \end{array}$$

with 1s in the unspecified places on the diagonal, 0s in all other places, and $1 \leqq i < j \leqq n$. Then γ_{ij} is a differentiable parametrized curve in $O^+(n)$. The tangent vector to γ_{ij} at $t = 0$ is

$$\alpha_{ij} = \left(\frac{\partial}{\partial x_{ji}}\right)_{\varepsilon} - \left(\frac{\partial}{\partial x_{ij}}\right)_{\varepsilon}$$

Clearly the $\frac{1}{2}n(n-1)$ vectors α_{ij} are linearly independent, which completes the proof.

EXERCISE

Prove that the set $\mathfrak{S}$ consisting of all $n \times n$ matrices of the form

$$\begin{pmatrix} 1 & a_{12} & \cdots & a_{1n} \\ 0 & 1 & \cdots & a_{2n} \\ \cdot & \cdot & \cdots & \cdot \\ 0 & 0 & \cdots & 1 \end{pmatrix}$$

is a Lie group and that the exponential mapping in $\mathfrak{S}$ maps the tangent space at the identity element in a one-to-one manner globally onto $\mathfrak{S}$.

CHAPTER 8

Integral Manifolds and Lie Subgroups

8-1 A SURVEY OF THE PROBLEM

In this chapter the results of Chapter 7 for one-parameter subgroups of a Lie group will be generalized to higher-dimensional subgroups. We shall give up the requirement that the subgroup be parametrized. Consequently it will be necessary to weaken the notions of left-invariant vector field and integral manifold.

Let $\mathfrak{G}$ be a Lie group. A subgroup of $\mathfrak{G}$ which is also a submanifold of $\mathfrak{G}$ will be called a Lie subgroup of $\mathfrak{G}$. We shall find that the Lie subgroups of a Lie group $\mathfrak{G}$ correspond to certain subspaces of the tangent space to $\mathfrak{G}$ at the identity element.

Let $\mathfrak{H}$ be a Lie subgroup of $\mathfrak{G}$ of dimension greater than zero. The cosets $g\mathfrak{H}$ with $g \in \mathfrak{G}$ constitute a family of submanifolds of $\mathfrak{G}$, since, for a fixed element g, the mapping $h \to gh$ of $\mathfrak{H}$ into $\mathfrak{G}$ is a diffeomorphism. In this chapter the manifolds $g\mathfrak{H}$ take the place of the integral manifolds that were studied in Chapter 7. The tangent vector to an integral manifold must be replaced here by the tangent space $X(g)$ to the submanifold $g\mathfrak{H}$ at the point g. $X(g)$ may be regarded as a subspace of the tangent space to $\mathfrak{G}$ at g. Thus we must replace our former notion of vector field by a correspondence X which associates with each point g of $\mathfrak{G}$ a subspace $X(g)$ of the tangent space to $\mathfrak{G}$ at g.

The new kind of vector field X may also be described in the following way. Let $X_0 = X(e)$, where e is the identity element of $\mathfrak{G}$. Let $\sigma^g\colon \mathfrak{G} \to \mathfrak{G}$ be defined by $\sigma^g(g') = gg'$, and let $\sigma^g{}_*$ be the linear transformation that is induced by σ^g on the tangent space. Then

$$X(g) = \sigma^g{}_*(X_0)$$

It follows that X is left-invariant in the sense that

$$\sigma^g{}_*(X(g')) = X(\sigma_g(g'))$$

for g, $g' \in \mathfrak{G}$. Conversely, any left-invariant correspondence X of this new type must satisfy $\mathrm{X}(g) = \sigma^g{}_*(\mathrm{X}(e))$ for $g \in \mathfrak{G}$. Hence our new kinds of left-invariant vector fields are in one-to-one correspondence with the subspaces of the tangent space to $\mathfrak{G}$ at e.

The meaning of integral manifold which we shall adopt is the following: A submanifold S of $\mathfrak{G}$ is an integral manifold for X if at each point $s \in S$ the tangent space to S is $\mathrm{X}(s)$. In this sense, the cosets $g\mathfrak{H}$ of the Lie subgroup $\mathfrak{H}$ are integral manifolds for the left-invariant vector field X which is derived from the tangent space to $\mathfrak{H}$ at the identity element.

The generalized ideas of vector field and integral manifold which have just been sketched are not confined to Lie groups. We shall first study them on any differentiable manifold. Before we do so, we shall present an example which shows that not every left-invariant vector field on a Lie group $\mathfrak{G}$ may be derived from a Lie subgroup. We therefore face the question of how to characterize those subspaces of the tangent space to $\mathfrak{G}$ at the identity element which are tangent spaces to Lie subgroups of $\mathfrak{G}$. We shall eventually answer this question.

Let $\mathfrak{G}$ be the Lie group which consists of all 3×3 matrices of the form

$$\begin{pmatrix} 1 & x_1 & x_2 \\ 0 & 1 & x_3 \\ 0 & 0 & 1 \end{pmatrix}$$

The tangent space to $\mathfrak{G}$ at the identity element consists of all matrices of the form

$$\begin{pmatrix} 0 & a_1 & a_2 \\ 0 & 0 & a_3 \\ 0 & 0 & 0 \end{pmatrix}$$

Let $\mathrm{X}(e)$ be the subspace which consists of all matrices of the form

$$\begin{pmatrix} 0 & a_1 & 0 \\ 0 & 0 & a_3 \\ 0 & 0 & 0 \end{pmatrix}$$

Assume that $\mathfrak{H}$ is a Lie subgroup of $\mathfrak{G}$ and that the tangent space to $\mathfrak{H}$ is $\mathrm{X}(e)$. Then $\mathfrak{H}$ contains the one-parameter subgroups

$$\exp\begin{pmatrix} 0 & t & 0 \\ 0 & 0 & 0 \\ 0 & 0 & 0 \end{pmatrix} \qquad \exp\begin{pmatrix} 0 & 0 & 0 \\ 0 & 0 & t \\ 0 & 0 & 0 \end{pmatrix}$$

and hence all matrices of the forms

$$\begin{pmatrix} 1 & t & 0 \\ 0 & 1 & 0 \\ 0 & 0 & 1 \end{pmatrix} \qquad \begin{pmatrix} 1 & 0 & 0 \\ 0 & 1 & t \\ 0 & 0 & 1 \end{pmatrix}$$

In particular, $\mathfrak{H}$ contains the matrices

$$\begin{pmatrix} 1 & u & 0 \\ 0 & 1 & 0 \\ 0 & 0 & 1 \end{pmatrix}\begin{pmatrix} 1 & 0 & 0 \\ 0 & 1 & u \\ 0 & 0 & 1 \end{pmatrix}\begin{pmatrix} 1 & -u & 0 \\ 0 & 1 & 0 \\ 0 & 0 & 1 \end{pmatrix}\begin{pmatrix} 1 & 0 & 0 \\ 0 & 1 & -u \\ 0 & 0 & 1 \end{pmatrix} = \begin{pmatrix} 1 & 0 & u^2 \\ 0 & 1 & 0 \\ 0 & 0 & 1 \end{pmatrix}$$

and hence the one-parameter subgroup

$$\begin{pmatrix} 1 & 0 & t \\ 0 & 1 & 0 \\ 0 & 0 & 1 \end{pmatrix}$$

But the tangent vector to this curve at the origin is

$$\begin{pmatrix} 0 & 0 & 1 \\ 0 & 0 & 0 \\ 0 & 0 & 0 \end{pmatrix}$$

which is not in $X(e)$. Thus our assumption that $\mathfrak{H}$ exists is incorrect, as we wished to demonstrate.

8-2 p-VECTOR FIELDS AND INTEGRAL MANIFOLDS

Let M be a differentiable manifold. Let U be an open subset of M. A correspondence X which assigns to each point m of U a subspace $X(m)$ of dimension p of the tangent space to M at m, in such a way that p is the same for all points of U, is called a *p-vector field* on U.

Let m° be a point of U. If there is a neighborhood U_1 of m° and differentiable vector fields $\chi_1, \ldots, \chi_p$ on U_1 such that for each point m in U_1 the vector space $X(m)$ is spanned by $\chi_1(m), \ldots, \chi_p(m)$, then X is said to be *differentiable* at m°. Under these circumstances, the vector fields $\chi_1, \ldots, \chi_p$ are linearly independent on U_1. X is differentiable on U if it is differentiable at each point of U.

The p-vector field X will be called *integrable* at the point m° if there is a coordinate neighborhood U_1 containing m°, with coordinates $(x_1, \ldots, x_n)$, such that for each point m of U_1 the vector space $X(m)$ is spanned by the vectors

$$\left(\frac{\partial}{\partial x_1}\right)_m, \ldots, \left(\frac{\partial}{\partial x_p}\right)_m$$

Under these circumstances, if $m = (c_1, \ldots, c_n)$ is a point of U_1, the submanifold of U_1 defined by the equations

$$x_{p+1} = c_{p+1}, \ldots, x_n = c_n$$

is an integral manifold for X which contains m. The following theorem shows that this integral manifold is essentially unique.

Theorem 8-1. *Let U be a coordinate neighborhood of a manifold with coordinates $(x_1, \ldots, x_n)$. Let X be the p-vector field on U which is spanned by the vector fields*

$$\frac{\partial}{\partial x_1}, \ldots, \frac{\partial}{\partial x_p}$$

If N is an integral manifold for X which contains the point $(0, \ldots, 0) \in U$ and if every point of N may be joined to $(0, \ldots, 0)$ by a differentiable parametrized curve which lies in N, then N is an open subset of the submanifold of U defined by the equations $x_{p+1} = \cdots = x_n = 0$.

Proof. Let $m = (c_1, \ldots, c_n)$ be a point of N. We must prove that $c_{p+1} = \cdots = c_n = 0$. Let γ be a differentiable curve in N which joins $(0, \ldots, 0)$ to m. We may write

$$\gamma(t) = (g_1(t), \ldots, g_n(t))$$

It suffices to prove that $g_{p+1}, \ldots, g_n$ are zero functions. But the tangent vector to γ at $\gamma(t)$ must lie in X$(\gamma(t))$. Hence

$$g'_{p+1} = \cdots = g'_n = 0$$

It follows that $g_{p+1}, \ldots, g_n$ are constant. Since

$$g_{p+1}(0) = \cdots = g_n(0) = 0$$

the theorem is proved.

Corollary. *Let χ be a vector field and X a p-vector field, both of which are defined on an open subset U of a differentiable manifold. Let $\chi(m) \in$ X(m) for $m \in U$. Let γ be an integral manifold for χ and N an integral manifold for X with $\gamma(0) \in N$. If X is integrable at $\gamma(0)$, then there is an $\epsilon > 0$ such that $\gamma(t) \in N$ for $|t| < \epsilon$.*

EXERCISES

1. Prove the corollary.

2. Let S^3 be the 3-sphere, that is, the differentiable variety in E^4 defined by the equation

$$x_1^2 + x_2^2 + x_3^2 + x_4^2 = 1$$

Consider the vector fields

$$\chi_1 = (1 - x_4 - x_1^2)\frac{\partial}{\partial x_1} - x_1x_2\frac{\partial}{\partial x_2} - x_1x_3\frac{\partial}{\partial x_3} + x_1(1 - x_4)\frac{\partial}{\partial x_4}$$

$$\chi_2 = -x_1x_2\frac{\partial}{\partial x_1} + (1 - x_4 - x_2^2)\frac{\partial}{\partial x_2} - x_2x_3\frac{\partial}{\partial x_3} + x_2(1 - x_4)\frac{\partial}{\partial x_4}$$

Let S^* be the set of points of S^3 excepting (0, 0, 0, 1).

a. Show that χ_1 and χ_2 are linearly independent vector fields on S^*.

Let X be the 2-vector field on S^* spanned by χ_1 and χ_2.

b. Show that the 2-dimensional differentiable varieties defined by the relations

$$\begin{aligned} x_3 + kx_4 &= k \\ x_1^2 + x_2^2 + x_3^2 + x_4^2 &= 1 \\ kx_3 - x_4 &\neq 0 \\ x_4 &\neq 1 \end{aligned}$$

$-\infty < k < \infty$, are integral manifolds for X.

8-3 THE INTEGRABILITY OF VECTOR FIELDS

In order to study the integrability of p-vector fields, we need information concerning the manner in which the solution of a system of differential equations depends upon the boundary values. The Picard theorem as it is stated in Section 4-4 will suffice for our purposes.

One use of the Picard theorem in the theory of differentiable manifolds may be illustrated by the following situation. Let U and V be coordinate neighborhoods of two differentiable manifolds. Let $\sigma\colon V \to U$ be a differentiable mapping. Let χ be a differentiable vector field on U. We shall see that it is possible to extend σ to a differentiable mapping $\gamma\colon V' \times J \to U$, where J is a neighborhood of 0 in E^1 and V' is an open subset of V, in such a way that $\gamma(y, t)$ is an integral manifold for χ when y is a fixed point in V' and $\gamma(y, 0) = \sigma(y)$ for $y \in V'$. Furthermore, if y° is a point of V, then V' may be required to contain y°.

Let $(x_1, \ldots, x_n)$ be coordinates in U. Then χ determines a system of first-order differential equations

$$\frac{dx_1}{dt} = a_1(x_1, \ldots, x_n)$$

$$\cdots\cdots\cdots\cdots$$

$$\frac{dx_n}{dt} = a_n(x_1, \ldots, x_n)$$

where

$$\chi = a_1\frac{\partial}{\partial x_1} + \cdots + a_n\frac{\partial}{\partial x_n}$$

whose solutions are integral manifolds for χ. Let $y^\circ \in V$ and $x^\circ = \sigma(y^\circ)$. By the Picard theorem, there are an $\epsilon > 0$, a neighborhood U° of x°, and integral manifolds $\Gamma(x, t)$ for χ for each $x \in U^\circ$ and $|t| < \epsilon$ such that $x = \Gamma(x, 0)$ and the mapping $(x, t) \to \Gamma(x, t)$ is differentiable. Let $V' = \sigma^{-1}(U^\circ)$, let J be the interval $-\epsilon < t < \epsilon$, and define $\gamma(y, t) = \Gamma(\sigma(y), t)$.

We shall apply the Picard theorem in this manner in the following theorem.

Theorem 8-2. *Let χ be a differentiable vector field on an open subset U of a differentiable manifold M. If $\chi(m^\circ) \neq 0$, then there is a coordinate neighborhood U_1 of m° with coordinates $(x_1, \ldots, x_n)$ such that $\chi = \partial/\partial x_1$ on U_1.*

Proof. Let W be a coordinate neighborhood of m° with coordinates $(y_1, \ldots, y_n)$. We may assume that $m^\circ = (0, \ldots, 0)$ and

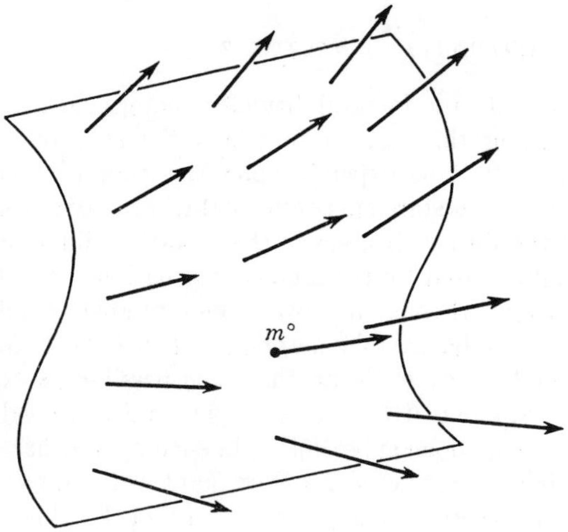

FIG. 8-1

$(\partial/\partial y_1)_{m^\circ} = \chi(m^\circ)$. We may also assume that W is sufficiently small so that $\chi(m) \neq 0$ for $m \in W$. Let V be the open subset of E^{n-1} consisting of those points $(y_2, \ldots, y_n)$ such that $(0, y_2, \ldots, y_n) \in W$. Define $\sigma: V \to W$ by $\sigma(y_2, \ldots, y_n) = (0, y_2, \ldots, y_n)$.

Figure 8-1 illustrates this situation for $n = 3$. It shows the surface $\sigma(V)$ along which $y_1 = 0$ with the vectors of the vector field χ attached to it. Only the vector at m° need have the direction $\partial/\partial y_1$.

If we apply the Picard theorem as we have described, there are a neighborhood J of 0 in E^1 and a neighborhood V' of $(0, \ldots, 0)$ in V and a differentiable mapping $\gamma: J \times V' \to W$ such that $\gamma(t, x_2, \ldots, x_n)$ is an integral manifold for χ when $x_2, \ldots, x_n$ are fixed and

$\gamma(0, x_2, \ldots, x_n) = (0, x_2, \ldots, x_n)$. We shall show that γ defines a coordinate neighborhood of M with the desired properties.

Let $(y_1, \ldots, y_n) = \gamma(x_1, \ldots, x_n)$. At the point $(0, \ldots, 0)$

$$\begin{pmatrix} \dfrac{\partial y_1}{\partial x_1} & \cdots & \dfrac{\partial y_1}{\partial x_n} \\ \cdots & \cdots & \cdots \\ \dfrac{\partial y_n}{\partial x_1} & \cdots & \dfrac{\partial y_n}{\partial x_n} \end{pmatrix} = \begin{pmatrix} 1 & 0 & \cdots & 0 \\ 0 & 1 & \cdots & 0 \\ \cdots & \cdots & \cdots & \cdots \\ 0 & 0 & \cdots & 1 \end{pmatrix}$$

Consequently γ is regular at $(0, \ldots, 0)$, and there is a neighborhood of this point on which γ is a diffeomorphism. Thus $(x_1, \ldots, x_n)$ become coordinates on this neighborhood. With respect to these coordinates, $\partial/\partial x_1 = \chi$, since $\gamma(x_1, \ldots, x_n)$ is an integral manifold for χ as a function of x_1.

In brief, Theorem 8-2 says that a differentiable vector field is integrable at each point at which it does not vanish.

8-4 LIE DERIVATIVES OF VECTOR FIELDS

Let φ be a differentiable vector field on an open subset U of a differentiable manifold M. If $\varphi \neq 0$ on U, then φ determines a flow on U which has no singularities. We shall show how to measure the rate of change along this flow of any other differentiable vector field on U.

Let $m^\circ \in U$. By applying the reformulation of the Picard theorem given in the previous section, with σ the identity mapping, there are a neighborhood J of 0 in E^1, a neighborhood U' of m°, and a differentiable mapping γ: $U' \times J \to U$ such that $\gamma(m, t)$ is an integral manifold for φ as a function of t and $\gamma(m, 0) = m$. Define $\mu_t(m) = \gamma(m, t)$. For each $t \in J$ we may regard μ_t: $U' \to U$ as a kind of displacement of U' along integral manifolds for φ.

We shall prove that, if the interval J and the neighborhood U' are sufficiently small, the displacements μ_t are all diffeomorphisms. By Theorem 8-2, we may assume that U' is a coordinate neighborhood with coordinates $(x_1, \ldots, x_n)$ such that $\partial/\partial x_1 = \varphi$. The mapping γ: $U' \times J \to U$ is continuous, and so $\gamma^{-1}(U')$ is an open subset of $U' \times J$ containing $(m^\circ, 0)$. Hence there are a neighborhood U'' of m° contained in U' and a neighborhood J' of 0 contained in J such that $\gamma(U'' \times J') \subset U'$. We may conclude that there is an $\epsilon > 0$ such that $|t| < \epsilon$ implies $\mu_t(U'') \subset U'$. For $|t| < \epsilon$ and $(x_1, \ldots,$

$x_n) \in U''$, we then have

$$\mu_t(x_1, \ldots, x_n) = (x_1 + t, x_2, \ldots, x_n)$$

which is clearly a diffeomorphism. We shall call the family $\{\mu_t\}$ of diffeomorphisms the *one-parameter family of motions determined by* φ.

For m near m°, let λ_t: $T(\mu_t(m)) \to T(m)$ be the linear transformation which is induced on the tangent space by the *inverse* of the diffeomorphism μ_t. Let χ be a differentiable vector field on a neighborhood of m°. We may use λ_t to move the vector $\chi(\mu_t(m))$ to the point m so that it may be compared with the vector $\chi(m)$. Define

$$\Delta_\varphi \chi(m) = \lim_{t \to 0} \frac{1}{t} [\lambda_t\{\chi(\mu_t(m))\} - \chi(m)]$$

for m near m°. We shall show that this derivative exists and consequently defines a vector field $\Delta_\varphi \chi$ on a neighborhood of m°.

In terms of the coordinates in U',

$$\chi = a_1 \frac{\partial}{\partial x_1} + \cdots + a_n \frac{\partial}{\partial x_n}$$

where the functions $a_1, \ldots, a_n$ are differentiable near m°. Since

$$\mu_t(x_1, \ldots, x_n) = (x_1 + t, x_2, \ldots, x_n)$$

it follows that

$$\Delta_\varphi \chi(m) = \left.\frac{\partial a_1}{\partial x_1}\right|_m \left(\frac{\partial}{\partial x_1}\right)_m + \cdots + \left.\frac{\partial a_n}{\partial x_1}\right|_m \left(\frac{\partial}{\partial x_n}\right)_m$$

Thus $\Delta_\varphi \chi$ is a differentiable vector field. $\Delta_\varphi \chi$ is called the *Lie derivative* of χ with respect to φ.

EXERCISE

Let χ be a differentiable vector field on an open subset U of a manifold, and let f be a differentiable function on U. Let φ be a nonzero differentiable vector field on U. Let $\psi(m) = f(m)\chi(m)$ for $m \in U$. Prove that

$$\Delta_\varphi \psi = \langle \varphi, df \rangle \chi + f \Delta_\varphi \chi$$

on U.

8-5 ANOTHER FORMULATION OF THE LIE DERIVATIVE

Let φ and χ be differentiable vector fields on an open subset U of a differentiable manifold M. Let $m \in U$, and let $G^*(m)$ be the space of germs of differentiable functions at m. If $\kappa \in G^*(m)$ and if κ is represented by the differentiable function f, then $g = \langle \chi, df \rangle$ is a differentiable function on a neighborhood of m. Hence we may form the differentiable function $\langle \varphi, dg \rangle$ on a neighborhood of m. The mapping

$$\kappa \to \langle \varphi(m), (dg)_m \rangle$$

is clearly well-defined and is a linear functional on $G^*(m)$.

Similarly, we may form the function $h = \langle \varphi, df \rangle$ and the linear functional

$$\kappa \to \langle \chi(m), (dh)_m \rangle$$

We shall denote the linear functional

$$\kappa \to \langle \varphi(m), (dg)_m \rangle - \langle \chi(m), (dh)_m \rangle$$

on $G^*(m)$ by $[\varphi, \chi](m)$. In spite of the fact that the terms $\langle \varphi(m), (dg)_m \rangle$ and $\langle \chi(m), (dh)_m \rangle$ may be different for two germs whose differentials are the same, we shall prove that this is not the case for $[\varphi, \chi](m)$. Consequently $[\varphi, \chi](m)$ is a linear functional on the cotangent space $T^*(m)$, and so it represents a tangent vector to M at m.

Let U_1 be a coordinate neighborhood of m which is contained in U. Let $(x_1, \ldots, x_n)$ be the coordinates on U_1. Then $\varphi = \Sigma_{i=1}^n a_i(\partial/\partial x_i)$ and $\chi = \Sigma_{i=1}^n b_i(\partial/\partial x_i)$, where $a_1, \ldots, a_n, b_1, \ldots, b_n$ are functions. Hence $g = \Sigma_{i=1}^n b_i(\partial f/\partial x_i)$ and $h = \Sigma_{i=1}^n a_i(\partial f/\partial x_i)$, and so

$$\langle \varphi, dg \rangle = \sum_{i=1}^n a_i \frac{\partial}{\partial x_i} \left(\sum_{j=1}^n b_j \frac{\partial f}{\partial x_j} \right)$$

$$\langle \chi, dh \rangle = \sum_{i=1}^n b_i \frac{\partial}{\partial x_i} \left(\sum_{j=1}^n a_j \frac{\partial f}{\partial x_j} \right)$$

and finally

$$\langle \varphi, dg \rangle - \langle \chi, dh \rangle = \sum_{i=1}^n \sum_{j=1}^n \left\{ a_i \frac{\partial b_j}{\partial x_i} - b_i \frac{\partial a_j}{\partial x_i} \right\} \frac{\partial f}{\partial x_j}$$

It follows that if f_1 and f_2 represent germs with the same differential, which means $(df_1)_m = (df_2)_m$, the value of $[\varphi, \chi](m)$ is the same for both, as we asserted. $[\varphi, \chi]$ is therefore the vector field on U_1 which is given by the formula

$$[\varphi, \chi] = \sum_{j=1}^n \sum_{i=1}^n \left(a_i \frac{\partial b_j}{\partial x_i} - b_i \frac{\partial a_j}{\partial x_i} \right) \frac{\partial}{\partial x_j}$$

Theorem 8-3. *If φ and χ are differentiable vector fields on an open subset U of a differentiable manifold and if φ is nonzero on U, then*

$$\Delta_\varphi \chi = [\varphi, \chi]$$

on U.

Proof. Let $m \in U$. By Theorem 8-2, there is a coordinate neighborhood of m with coordinates $(x_1, \ldots, x_n)$ such that $\varphi = \partial/\partial x_1$. If $\chi = \Sigma_{i=1}^n b_i(\partial/\partial x_i)$, then

$$\Delta_\varphi \chi = \sum_{i=1}^n \frac{\partial b_i}{\partial x_1} \frac{\partial}{\partial x_i}$$

and also
$$[\varphi, \chi] = \sum_{i=1}^n \frac{\partial b_i}{\partial x_1} \frac{\partial}{\partial x_i}$$

According to Theorem 8-3, we may use the vector field $[\varphi, \chi]$ to extend the definition of the Lie derivative Δ_φ to the case where φ is not necessarily different from zero.

Corollary. *If φ and χ are differentiable vector fields on an open subset of a differentiable manifold, then*

$$\Delta_\varphi \chi = -\Delta_\chi \varphi$$

Corollary. *Let X be a p-vector field which is integrable on an open subset U of a differentiable manifold M. If φ and χ are differentiable vector fields on U such that $\varphi(m) \in X(m)$ and $\chi(m) \in X(m)$ for $m \in U$, then $\Delta_\varphi \chi(m) \in X(m)$ for $m \in U$.*

Proof. Let $m \in U$. Since X is integrable at m, there is a coordinate neighborhood U_1 of m with the coordinates $x_1, \ldots, x_n$ such that X is spanned by $\partial/\partial x_1, \ldots, \partial/\partial x_p$ on U_1. Thus

$$\varphi = \sum_{j=1}^p a_j \frac{\partial}{\partial x_j} \qquad \chi = \sum_{j=1}^p b_j \frac{\partial}{\partial x_j}$$

on U_1. By Theorem 8-3,

$$\Delta_\varphi \chi = \sum_{i=1}^p \sum_{j=1}^p \left(a_i \frac{\partial b_j}{\partial x_i} - b_i \frac{\partial a_j}{\partial x_i} \right) \frac{\partial}{\partial x_j}$$

and so

$$\Delta_\varphi \chi(m) \in X(m)$$

From this corollary we may derive a necessary condition for the integrability of a differentiable p-vector field. Let X be a p-vector field which is differentiable on an open subset U of a manifold M. Let $m^\circ \in U$. Then there are differentiable vector fields $\chi_1, \ldots, \chi_p$ such that at each point m of a neighborhood U_1 of m° the vectors $\chi_1(m), \ldots, \chi_p(m)$ span $X(m)$. Since these vectors are linearly independent, they are all different from zero, and we may form the Lie derivatives $\Delta_{\chi_i}\chi_j$ for $i, j = 1, \ldots, p$. *If* X *is integrable on* U, *then each of the Lie derivatives* $\Delta_{\chi_i}\chi_j(m)$ *is in* $X(m)$ *and is therefore a linear combination of the vectors* $\chi_1(m), \ldots, \chi_p(m)$ *for* $m \in U_1$.

Let $\chi_1, \ldots, \chi_p$ be linearly independent differentiable vector fields on an open subset U of a differentiable manifold, and let them span the p-vector field X. If on U there are functions $a_k^{i,j}$ such that

$$\Delta_{\chi_i}\chi_j = a_1^{i,j}\chi_1 + \cdots + a_p^{i,j}\chi_p$$

for $i, j = 1, \ldots, p$, then X will be said to be *closed* with respect to $\chi_1, \ldots, \chi_p$ on U.

In terms of this terminology, we have proved that if a differentiable p-vector field X is integrable at a point m, then X is closed with respect to any set of p differentiable vector fields which span it on some neighborhood of m.

EXERCISE

If χ, φ, and ψ are differentiable vector fields on an open subset of a differentiable manifold, prove the Jacobi identity

$$[\chi, [\varphi, \psi]] + [\varphi, [\psi, \chi]] + [\psi, [\chi, \varphi]] = 0$$

8-6 THE EXISTENCE OF INTEGRAL MANIFOLDS

In Section 8-5 we showed that a necessary condition for a p-vector field X spanned by the differentiable vector fields $\chi_1, \ldots, \chi_p$ to be integrable is that X be closed with respect to $\chi_1, \ldots, \chi_p$. In Section 8-7 we shall prove that this condition is also sufficient. Before presenting the proof, we shall examine the case of a 2-vector field in E^3 where the curves and surfaces under consideration are easily visualized.

Let χ_1 and χ_2 be linearly independent vector fields on an open subset U of E^3. Let χ_1 and χ_2 span the 2-vector field X. Let X be closed with respect to χ_1 and χ_2. How shall we construct an integral manifold for X? We may be guided by the fact that, if such an integral manifold exists, the integral manifolds for the vector fields χ_1 and χ_2 must lie in the integral manifold for X, at least locally.

Let m° be a point in E^3. Let γ be an integral manifold for χ_1

through m°. Through each point of γ, consider the integral manifold for χ_2. These curves determine a surface N in E^3, as illustrated in Figure 8-2. Consider the one-parameter family of motions $\{\mu_t\}$ determined by χ_2 on a neighborhood of m°. These are displacements of the surface N along itself in the direction of the integral manifolds for χ_2. In particular, the curve γ is displaced onto the curve $\mu_t(\gamma)$ by the diffeomorphism μ_t.

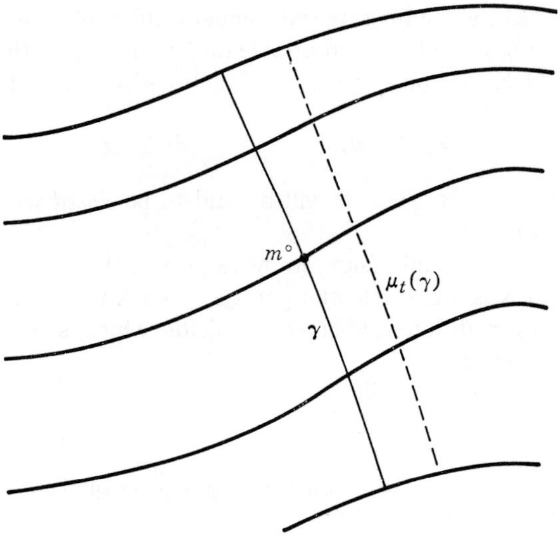

FIG. 8-2

Let ψ be the vector field on N which assigns to each point m the tangent vector to the curve $\mu_t(\gamma)$ at m. Then the vector fields χ_2 and ψ are linearly independent on N, and at each point m the vectors $\chi_2(m)$ and $\psi(m)$ constitute a basis of the tangent space to N at m. In order to show that N is an integral manifold for X, we must prove that ψ is a linear combination of χ_1 and χ_2.

χ_1 and χ_2 can be completed by a third differentiable vector field χ_3 in such a way that χ_1, χ_2, and χ_3 are linearly independent near m° and the Lie derivative of χ_3 with respect to χ_2 is zero. We may write

$$\psi = f_1\chi_1 + f_2\chi_2 + f_3\chi_3$$

where f_1, f_2, and f_3 are functions on N. The Lie derivatives of ψ, χ_2, and χ_3 with respect to χ_2 are all zero. In terms of the parametrization

$(s, t) \to \mu_t(\gamma(s))$ of the surface N, we have

$$0 = \frac{\partial f_1}{\partial t} \chi_1 + f_1 \Delta_{\chi_2}\chi_1 + \frac{\partial f_2}{\partial t} \chi_2 + \frac{\partial f_3}{\partial t} \chi_3$$

We now apply the assumption that $\Delta_{\chi_2}\chi_1$ is a linear combination of χ_1 and χ_2 to conclude that $\partial f_3/\partial t = 0$ on N. If we make sure that N is a connected set, then f_3 is a function of s alone. Since $f_3 = 0$ along $\gamma(s)$, it follows that $f_3 = 0$ on N, which concludes the proof that N is an integral manifold for X near m.

It is not difficult now to introduce coordinates (s, t, u) near $m°$ in such a way that $u = 0$ is the surface N, and this would complete the proof that X is integrable at $m°$. The methods that we have employed in this special case can be extended by induction to cover the general case. But the analytic proof that we shall give in the next section is much simpler, and so we shall not carry out the geometric proof.

EXERCISE

Fill in the missing details of this section.

8-7 THE INTEGRABILITY THEOREM

Theorem 8-4. *Let X be the p-vector field which is spanned by the linearly independent differentiable vector fields $\chi_1, \ldots, \chi_p$ on an open subset U of a differentiable manifold M. If X is closed with respect to $\chi_1, \ldots, \chi_p$, then X is integrable on U.*

Proof. The case $p = 1$ has been proved in Theorem 8-2. We shall prove the general case by induction on p. Assume that the theorem has been proved for $(p - 1)$-vector fields, $p \geqq 2$.

Let $m°$ be a point of U. There is a coordinate neighborhood U_1 of $m°$ with coordinates $(x_1, \ldots, x_n)$ such that $\chi_1 = \partial/\partial x_1$. We may assume that U_1 is contained in U and that $m° = (0, \ldots, 0)$. Then for $i = 2, \ldots, p$

$$\chi_i = \sum_{j=1}^{n} a_{ij} \frac{\partial}{\partial x_j}$$

where the a_{ij} are differentiable functions on U_1. Define

$$\psi_i = \chi_i - a_{i1}\chi_1 = \chi_i - a_{i1} \frac{\partial}{\partial x_1}$$

for $i = 2, \ldots, p$. Then $\psi_2, \ldots, \psi_p$ are linearly independent differentiable vector fields on U_1. Let Ψ be the $(p-1)$-vector field which is spanned by $\psi_2, \ldots, \psi_p$. In order to apply the induction hypothesis to Ψ, we must prove that Ψ is closed with respect to $\psi_2, \ldots, \psi_p$.

Let $\Omega(m)$ be the vector space spanned by

$$\left(\frac{\partial}{\partial x_2}\right)_m, \ldots, \left(\frac{\partial}{\partial x_n}\right)_m$$

for each point m in U_1. Then

$$\Psi(m) = X(m) \cap \Omega(m)$$

Since the Lie derivatives $\Delta_{\psi_i}\psi_j(m)$ are in $\Omega(m)$, it suffices to prove that they are in $X(m)$.

We have

$$\begin{aligned} \Delta_{\psi_i}\psi_j = [\psi_i, \psi_j] &= [\psi_i, \chi_j - a_{j1}\chi_1] \\ &= [\psi_i, \chi_j] - \langle \psi_i, da_{j1}\rangle \chi_1 - a_{j1}[\psi_i, \chi_1] \end{aligned}$$

and

$$\begin{aligned} [\psi_i, \chi_j] &= -[\chi_j, \psi_i] = -[\chi_j, \chi_i - a_{i1}\chi_1] \\ &= -[\chi_j, \chi_i] + \langle \chi_j, da_{i1}\rangle \chi_1 + a_{i1}[\chi_j, \chi_1] \end{aligned}$$

Since X is closed with respect to $\chi_1, \ldots, \chi_p$, it follows that $[\psi_i, \chi_j](m)$ is in $X(m)$ for $j = 1, \ldots, p$ and hence that $\Delta_{\psi_i}\psi_j(m)$ is in $X(m)$. This proves that Ψ is closed with respect to $\psi_2, \ldots, \psi_p$.

Let S be the submanifold of U_1 defined by the equation $x_1 = 0$. Then Ψ may be regarded as a $(p-1)$-vector field on S. If we apply the induction hypothesis to Ψ, we may assert that there is a coordinate neighborhood V of S containing m°, with coordinates $(y_2, \ldots, y_n)$, such that Ψ is spanned on V by the vector fields

$$\frac{\partial}{\partial y_2}, \ldots, \frac{\partial}{\partial y_p}$$

Let the coordinate transformation $(0, x_2, \ldots, x_n) \to (y_2, \ldots, y_n)$ on $V \cap U_1$ be given by

$$\begin{aligned} y_2 &= g_2(x_2, \ldots, x_n) \\ &\cdots\cdots\cdots\cdots \\ y_n &= g_n(x_2, \ldots, x_n) \end{aligned}$$

Define

$$\begin{aligned} z_1 &= x_1 \\ z_2 &= g_2(x_2, \ldots, x_n) \\ &\cdots\cdots\cdots\cdots \\ z_n &= g_n(x_2, \ldots, x_n) \end{aligned}$$

for $(x_1, \ldots, x_n)$ in a neighborhood U_2 of m° in U_1 such that $U_2 \cap S \subset V$. We shall prove that $(z_1, \ldots, z_n)$ are coordinates in U_2 such that

$$\frac{\partial}{\partial z_1}, \ldots, \frac{\partial}{\partial z_p}$$

span X on U_2.

Since the mapping $(0, x_2, \ldots, x_n) \to (y_2, \ldots, y_n)$ is a diffeomorphism, it is clear that the mapping $(x_1, \ldots, x_n) \to (z_1, \ldots, z_n)$ is a diffeomorphism. Hence this mapping defines a coordinate transformation on U_2. There are differentiable functions c_{ij} on U_2 such that

$$\chi_j = \sum_{k=1}^{n} c_{jk} \frac{\partial}{\partial z_k}$$

for $j = 1, \ldots, p$. Since X is closed with respect to $\chi_1, \ldots, \chi_p$, there are differentiable functions b_{ij} on U_2 such that

$$\Delta_{\chi_1}\chi_j = \sum_{k=1}^{p} b_{jk}\chi_k = \sum_{k=1}^{p} \sum_{l=1}^{n} b_{jk}c_{kl} \frac{\partial}{\partial z_l}$$

for $j = 1, \ldots, p$. On the other hand, $\chi_1 = \partial/\partial z_1$ on U_2, so that

$$\Delta_{\chi_1}\chi_j = \sum_{l=1}^{n} \frac{\partial c_{jl}}{\partial z_1} \frac{\partial}{\partial z_l}$$

Thus the functions $c_{1l}, \ldots, c_{pl}$ are solutions of the system of homogeneous linear differential equations

$$\frac{\partial c_{jl}}{\partial z_1} = \sum_{k=1}^{p} b_{jk}c_{kl}$$

$j = 1, \ldots, p$, as functions of z_1. When $l > p$, we know that the functions $c_{1l}, \ldots, c_{pl}$ satisfy the boundary conditions

$$c_{1l} = \cdots = c_{pl} = 0$$

when $z_1 = 0$, since the vector fields

$$\frac{\partial}{\partial z_1}, \ldots, \frac{\partial}{\partial z_p}$$

span X on $U_2 \cap S$. We may assume that U_2 is a set of the form $\{(z_1, \ldots, z_n) | z_1^2 + \cdots + z_n^2 < r^2\}$. Then the uniqueness of the solution of our system of differential equations assures us that the functions $c_{1l}, \ldots, c_{pl}$ are zero everywhere on U_2 for $l > p$. Hence $\chi_1, \ldots, \chi_p$ are linear combinations of the vector fields

$$\frac{\partial}{\partial z_1}, \ldots, \frac{\partial}{\partial z_p}$$

on U_2, which was to be proved.

8-8 MAXIMAL INTEGRAL MANIFOLDS

In the preceding sections we have considered integral manifolds only in the neighborhood of a point. In this section we shall demonstrate the existence of integral manifolds in the large.

Lemma 8-1. *Let* X *be an integrable p-vector field on an open subset of a differentiable manifold* M. *Let* N_1 *and* N_2 *be two integral manifolds for* X. *If* m *is a point in* $N_1 \cap N_2$, *then there is an integral manifold* N *for* X *which contains* m *and which is an open submanifold of* N_1 *and of* N_2.

Proof. Let U be a coordinate neighborhood of m with coordinates $(x_1, \ldots, x_n)$ such that X is spanned by

$$\frac{\partial}{\partial x_1}, \ldots, \frac{\partial}{\partial x_p}$$

on U and $m = (0, \ldots, 0)$. Then $N_1 \cap U$ and $N_2 \cap U$ are open submanifolds of N_1 and N_2, respectively, and hence are integral manifolds for X.

Let N_1' be the set of all points of $N_1 \cap U$ which may be joined to m by a differentiable curve lying in $N_1 \cap U$. Then N_1' is an open set in the topology of $N_1 \cap U$ and so may be given the structure of an open submanifold of $N_1 \cap U$. Thus N_1' becomes an integral manifold for X. We may define N_2' similarly as an open submanifold of $N_2 \cap U$.

By Theorem 8-1 each of N_1' and N_2' is contained in the submanifold S of U defined by the equations $x_{p+1} = \cdots = x_n = 0$. Since N_1' and N_2' are open submanifolds of S, we can take $N = N_1' \cap N_2'$ and the lemma is proved.

Theorem 8-5. *Let* X *be an integrable p-vector field on a differentiable manifold* M. *If* m° *is any point of* M, *then there is an integral manifold* N *for* X *which contains* m° *and which satisfies the following conditions:*

(i) *Every point of* N *may be joined to* m° *by a differentiable curve which lies in* N.

(ii) N *is maximal, in the sense that* N *is not a submanifold of any larger integral manifold for* X *which satisfies condition* i.

(iii) *If N_1 is any integral manifold for* X *which satisfies condition* i, *then N_1 is an open submanifold of N.*

Proof. Let S be the point set which is the union of all integral manifolds for X which satisfy condition i. We give a topology to S by defining the open sets of S to be the unions of integral manifolds for X which are contained in S. That this is a topology follows from the preceding lemma. It is clear that S is a Hausdorff space.

We shall make S into a differentiable manifold by describing a suitable family of coordinate neighborhoods. Let N_α be an integral manifold for X which satisfies condition i. Let U be a coordinate neighborhood of N_α. Then U is an integral manifold for X. Thus U is an open subset of S, and the topology on U as a manifold is the same as the topology induced by the topology on S. We define U to be a coordinate neighborhood of S, using the same coordinate mapping as for N_α. We must show that our coordinate neighborhoods are related by differentiable coordinate transformations.

Let N_α and N_β be two integral manifolds for X which satisfy condition i. Let U_α be a coordinate neighborhood of N_α, and U_β a coordinate neighborhood of N_β, and let φ_α and φ_β be the corresponding coordinate mappings. If $U_\alpha \cap U_\beta$ is not empty, we must show that $\varphi_\beta \circ \varphi_\alpha^{-1}$ is differentiable.

U_α and U_β are integral manifolds for X. Let m be a point in $U_\alpha \cap U_\beta$. According to Lemma 8-1, there is an integral manifold U for X which is an open submanifold of U_α and of U_β. Let φ_α' be the restriction of φ_α to U and φ_β' be the restriction of φ_β to U. Then φ_α' makes U into a coordinate neighborhood U_α' of the manifold U, and φ_β' makes U into a coordinate neighborhood U_β' of the same manifold U. Thus $\varphi_\beta' \circ \varphi_\alpha'^{-1}$ is differentiable at m, and it follows that $\varphi_\beta \circ \varphi_\alpha^{-1}$ is differentiable at m, which was to be proved.

Let N denote the differentiable manifold which has been constructed from the set S. The proof will be completed if we can show that N is a submanifold of M, for the conditions i, ii, and iii then follow easily. However, the identity mapping of N into M is locally the identity mapping of some integral manifold N_α into M. Thus this identity mapping is a differentiable mapping which is regular, and so N is a submanifold of M.

Actually we have not completed the proof that N is a differentiable manifold for we have not shown that N is a separable topological space. The proof of this fact is difficult and will be omitted. It may be found in C. Chevalley, "Theory of Lie Groups," I (Princeton University Press, Princeton, N.J., 1946).

8-9 LIE ALGEBRAS

Let $\mathfrak{A}$ be a vector space of finite dimension over the real numbers. $\mathfrak{A}$ will be called a *Lie algebra* if there is an operation on $\mathfrak{A}$ which associates with each pair of elements α, β of $\mathfrak{A}$ an element $[\alpha, \beta]$ of $\mathfrak{A}$ in such a way that it satisfies the following axioms:

$$(a) \qquad \begin{aligned} [a_1\alpha_1 + a_2\alpha_2, \beta] &= a_1[\alpha_1, \beta] + a_2[\alpha_2, \beta] \\ [\alpha, a_1\beta_1 + a_2\beta_2] &= a_1[\alpha, \beta_1] + a_2[\alpha, \beta_2] \end{aligned}$$

for numbers a_1 and a_2 and $\alpha, \alpha_1, \alpha_2, \beta, \beta_1, \beta_2 \in \mathfrak{A}$.

$$(b) \qquad [\alpha, \alpha] = 0$$

for all $\alpha \in \mathfrak{A}$.

$$(c) \qquad [[\alpha, \beta], \gamma] + [[\beta, \gamma], \alpha] + [[\gamma, \alpha], \beta] = 0$$

for $\alpha, \beta, \gamma \in \mathfrak{A}$.

As an example of a Lie algebra, consider the set $L(n)$ of all $n \times n$ matrices with real components. Define

$$[\mathfrak{M}, \mathfrak{N}] = \mathfrak{M}\mathfrak{N} - \mathfrak{N}\mathfrak{M}$$

in terms of the usual operations among matrices. The reader may easily verify the axioms.

Another example, which is of basic importance in the theory of Lie groups, is the Lie algebra of a Lie group, which we shall now describe.

Lemma 8-2. *Let $\mathfrak{G}$ be a Lie group. Let φ and χ be two left-invariant vector fields on $\mathfrak{G}$. Then $[\varphi, \chi]$ is also left-invariant.*

Proof. We may assume that φ is different from zero everywhere on $\mathfrak{G}$, for otherwise φ is identically zero.

Let g_0 be an element of $\mathfrak{G}$. Let σ be the diffeomorphism $g \to g_0 g$, and let σ_* be the linear transformation induced on the tangent space at a point by σ. Since φ is left-invariant, $\sigma_* \circ \varphi = \varphi \circ \sigma$.

Let $\{\mu_t'\}$ be the one-parameter family of motions determined by φ on a neighborhood of $\sigma(g)$. Let $\{\mu_t\}$ be the one-parameter family of motions determined by φ on a neighborhood of g. If γ is an integral manifold for φ through g, then $\sigma \circ \gamma$ is an integral manifold for φ through $\sigma(g)$. Hence $\mu_t' \circ \sigma = \sigma \circ \mu_t$, and therefore $\lambda_t'^{-1} \circ \sigma_* = \sigma_* \circ \lambda_t^{-1}$, where λ_t' and λ_t are the linear transformations of the tangent spaces induced by the inverses of the diffeomorphisms μ_t' and μ_t, respectively.

Since χ is left-invariant, $\sigma_* \circ \chi = \chi \circ \sigma$. Hence

$$\begin{aligned}\{\sigma_* \circ [\varphi, \chi]\}(g) &= \lim_{t\to 0} \frac{1}{t} [(\sigma_* \circ \lambda_t)\{\chi(\mu_t(g))\} - \sigma_*\{\chi(g)\}] \\ &= \lim_{t\to 0} \frac{1}{t} [(\lambda_t' \circ \sigma_*)\{\chi(\mu_t(g))\} - \sigma_*\{\chi(g)\}] \\ &= \lim_{t\to 0} \frac{1}{t} [\lambda_t'\{\chi([\mu_t' \circ \sigma](g))\} - \chi(\sigma(g))] \\ &= [\varphi, \chi](\sigma(g))\end{aligned}$$

as was to be proved.

Let $\mathfrak{A}$ be the vector space of all left-invariant vector fields on $\mathfrak{G}$. The preceding lemma shows that we may introduce the operation in $\mathfrak{A}$ which assigns the product $[\varphi, \chi]$ to the pair of vector fields φ, χ from $\mathfrak{A}$. We then obtain a Lie algebra which will be called the Lie algebra of $\mathfrak{G}$ (see the exercise of Section 8-5).

It is easy to transfer the Lie-algebra structure of $\mathfrak{A}$ to the tangent space to $\mathfrak{G}$ at the identity element e. If α_1 and α_2 are two tangent vectors at e, they determine unique left-invariant vector fields χ_1 and χ_2 on $\mathfrak{G}$, respectively. Then we may define $[\alpha_1, \alpha_2] = [\chi_1, \chi_2](e)$. This makes the tangent space at e into a Lie algebra with the same structure as $\mathfrak{A}$.

Let $\mathfrak{A}$ be any Lie algebra. Then, from $[\alpha + \beta, \alpha + \beta] = 0$, we conclude that $[\alpha, \alpha + \beta] + [\beta, \alpha + \beta] = 0$ or

$$[\alpha, \beta] + [\beta, \alpha] = 0$$

This condition is equivalent to axiom *b*. For if $[\alpha, \beta] + [\beta, \alpha] = 0$, then, in particular, $2[\alpha, \alpha] = 0$ and so $[\alpha, \alpha] = 0$.

Let $\{\alpha_1, \ldots, \alpha_n\}$ be a basis of the vector space $\mathfrak{A}$. Then

$$[\alpha_i, \alpha_j] = \sum_{k=1}^{n} c_{ijk}\alpha_k$$

where c_{ijk} are numbers. The numbers c_{ijk} are called the *constants of structure* of $\mathfrak{A}$ with respect to the basis $\{\alpha_1, \ldots, \alpha_n\}$. Axioms *b* and *c* are reflected in certain relations among the constants of structure. From $[\alpha_i, \alpha_j] + [\alpha_j, \alpha_i] = 0$ we obtain

$$\sum_{k=1}^{n} c_{ijk}\alpha_k + \sum_{k=1}^{n} c_{jik}\alpha_k = 0$$

and thus

$$c_{ijk} + c_{jik} = 0 \tag{8-9-1}$$

From the Jacobi identity

$$[[\alpha_i, \alpha_j], \alpha_k] + [[\alpha_j, \alpha_k], \alpha_i] + [[\alpha_k, \alpha_i], \alpha_j] = 0$$

we obtain

$$\sum_{l=1}^{n} c_{ijl}[\alpha_l, \alpha_k] + \sum_{l=1}^{n} c_{jkl}[\alpha_l, \alpha_i] + \sum_{l=1}^{n} c_{kil}[\alpha_l, \alpha_j] = 0$$

Hence

$$\sum_{h=1}^{n}\sum_{l=1}^{n} c_{ijl}c_{lkh}\alpha_h + \sum_{h=1}^{n}\sum_{l=1}^{n} c_{jkl}c_{lih}\alpha_h + \sum_{h=1}^{n}\sum_{k=1}^{n} c_{kil}c_{ljh}\alpha_h = 0$$

so that

$$\sum_{l=1}^{n} (c_{ijl}c_{lkh} + c_{jkl}c_{lih} + c_{kil}c_{ljh}) = 0 \tag{8-9-2}$$

Conversely, let $\mathfrak{A}$ be any vector space of dimension n with the basis $\{\alpha_1, \ldots, \alpha_n\}$. Let $\{c_{ijk}\}$ be any set of numbers $1 \leqq i, j, k \leqq n$ which satisfy (8-9-1) and (8-9-2). Define

$$[\alpha_i, \alpha_j] = \sum_{k=1}^{n} c_{ijk}\alpha_k$$

and

$$\left[\sum_{i=1}^{n} a_i\alpha_i, \sum_{j=1}^{n} b_j\alpha_j\right] = \sum_{i=1}^{n}\sum_{j=1}^{n} a_ib_j[\alpha_i, \alpha_j]$$

Then axiom a is easily verified. Furthermore, condition (8-9-1) implies that $[\alpha_i, \alpha_j] + [\alpha_j, \alpha_i] = 0$, from which axiom b is easily derived. Axiom c may be obtained from (8-9-2), and thus $\mathfrak{A}$ is a Lie algebra.

We see from this result that the construction of all Lie algebras of dimension n is equivalent to the determination of all solutions of equations (8-9-1) and (8-9-2).

Let $\mathfrak{G}$ be a Lie group and $\mathfrak{A}$ the Lie algebra of $\mathfrak{G}$. If $\mathfrak{H}$ is a Lie subgroup of $\mathfrak{G}$ and if $\mathfrak{B}$ is the Lie algebra of $\mathfrak{H}$, then $\mathfrak{B}$ may be regarded as a subalgebra of $\mathfrak{A}$. Specifically, the elements of $\mathfrak{B}$ may be considered as elements of $\mathfrak{A}$, and when this is done the product of two elements of $\mathfrak{B}$ is the same in $\mathfrak{A}$ as it is in $\mathfrak{B}$.

Theorem 8-6. *Let $\mathfrak{A}$ be the Lie algebra of a Lie group $\mathfrak{G}$. If $\mathfrak{B}$ is a subalgebra of $\mathfrak{A}$, then there is a Lie subgroup $\mathfrak{H}$ of $\mathfrak{G}$ such that $\mathfrak{B}$ is the Lie algebra of $\mathfrak{H}$.*

Proof. As we have seen, we may regard $\mathfrak{B}$ as a subspace of the tangent space to $\mathfrak{G}$ at the identity element e. Consequently $\mathfrak{B}$ determines a unique left-invariant p-vector field X on $\mathfrak{G}$. Indeed, if $\{\beta_1, \ldots, \beta_p\}$ is a basis of the vector space $\mathfrak{B}$, the unique left-invariant vector fields $\chi_1, \ldots, \chi_p$ which are determined by $\beta_1, \ldots, \beta_p$, respectively, span X everywhere on $\mathfrak{G}$. Since $\mathfrak{B}$ is a subalgebra of $\mathfrak{A}$, it follows that $[\chi_i, \chi_j](e) = [\beta_i, \beta_j]$ is an element of $\mathfrak{B}$ and hence is a linear combination of $\beta_1, \ldots, \beta_p$:

$$[\chi_i, \chi_j](e) = a_1\beta_1 + \cdots + a_p\beta_p = a_1\chi_1(e) + \cdots + a_p\chi_p(e)$$

Since the vector field $[\chi_i, \chi_j]$ is also left-invariant,

$$[\chi_i, \chi_j] = a_1\chi_1 + \cdots + a_p\chi_p$$

everywhere on $\mathfrak{G}$; this proves that X is closed with respect to $\chi_1, \ldots, \chi_p$ on $\mathfrak{G}$. It follows that X is integrable on $\mathfrak{G}$.

Let $\mathfrak{H}$ be the maximal connected integral manifold for X which contains e. Let h be a point of $\mathfrak{H}$. Then $h^{-1}\mathfrak{H}$ is also a connected integral manifold for X which contains e. Hence $h^{-1}\mathfrak{H}$ is contained in $\mathfrak{H}$, and it follows that $\mathfrak{H}$ is a Lie subgroup of $\mathfrak{G}$. Since the tangent space to $\mathfrak{H}$ at e is $\mathfrak{B}$, it follows that $\mathfrak{B}$ is the Lie algebra of $\mathfrak{H}$.

This theorem answers the question that was raised at the beginning of this chapter. For it characterizes those subspaces of the tangent space to a Lie group $\mathfrak{G}$ at the identity element e (now possessing the structure of a Lie algebra) which are tangent spaces to Lie subgroups of $\mathfrak{G}$. They are precisely those subspaces which are also subalgebras under the Lie-algebra structure.

EXERCISE

Prove axioms a, b, and c from (8-9-1) and (8-9-2).

8-10 THE LIE ALGEBRA OF THE GENERAL LINEAR GROUP

Let $GL(n)$ be the general linear group on V^n. We have seen that the tangent space to $GL(n)$ at the identity element may be identified with the vector space of all $n \times n$ matrices. We now know that this tangent space has the structure of a Lie algebra, and the next theorem describes this structure in terms of the known operations on matrices.

Theorem 8-7. *If the $n \times n$ matrices $\mathfrak{M}$ and $\mathfrak{N}$ are regarded as tangent vectors to $GL(n)$ at the identity element, then*

$$[\mathfrak{M}, \mathfrak{N}] = \mathfrak{M}\mathfrak{N} - \mathfrak{N}\mathfrak{M}$$

Proof. Let $\mathfrak{M} = (a_{ij})$ and $\mathfrak{N} = (b_{ij})$. Let $\mathfrak{X} = (x_{ij})$ be an element of $GL(n)$. Let φ and χ be the left-invariant vector fields on $GL(n)$

determined by $\mathfrak{M}$ and $\mathfrak{N}$, respectively. Then

$$\varphi(\mathfrak{X}) = \sum_{i=1}^{n} \sum_{j=1}^{n} \sum_{k=1}^{n} x_{ij} a_{jk} \frac{\partial}{\partial x_{ik}}$$

$$\chi(\mathfrak{X}) = \sum_{i=1}^{n} \sum_{j=1}^{n} \sum_{k=1}^{n} x_{ij} b_{jk} \frac{\partial}{\partial x_{ik}}$$

Hence

$$\begin{aligned}
[\varphi, \chi](\mathfrak{X}) &= \sum_{i=1}^{n} \sum_{j=1}^{n} \sum_{k=1}^{n} \sum_{l=1}^{n} \left\{ \sum_{p=1}^{n} x_{ip} a_{pj} \frac{\partial}{\partial x_{ij}} \sum_{q=1}^{n} x_{kq} b_{ql} \right. \\
&\qquad \left. - \sum_{p=1}^{n} x_{ip} b_{pj} \frac{\partial}{\partial x_{ij}} \sum_{q=1}^{n} x_{kq} a_{ql} \right\} \frac{\partial}{\partial x_{kl}} \\
&= \sum_{i=1}^{n} \sum_{j=1}^{n} \sum_{l=1}^{n} \sum_{p=1}^{n} \{ x_{ip} a_{pj} b_{jl} - x_{ip} b_{pj} a_{jl} \} \frac{\partial}{\partial x_{il}} \\
&= \sum_{i=1}^{n} \sum_{p=1}^{n} \sum_{l=1}^{n} x_{ip} \left\{ \sum_{j=1}^{n} a_{pj} b_{jl} - \sum_{j=1}^{n} b_{pj} a_{jl} \right\} \frac{\partial}{\partial x_{il}}
\end{aligned}$$

On the other hand,

$$\mathfrak{M}\mathfrak{N} - \mathfrak{N}\mathfrak{M} = \left(\sum_{j=1}^{n} a_{ij} b_{jl} - \sum_{j=1}^{n} b_{ij} a_{jl} \right)$$

which shows that $[\varphi, \chi]$ is the left-invariant vector field determined by $\mathfrak{M}\mathfrak{N} - \mathfrak{N}\mathfrak{M}$. Consequently,

$$[\mathfrak{M}, \mathfrak{N}] = \mathfrak{M}\mathfrak{N} - \mathfrak{N}\mathfrak{M}$$

EXERCISES

1. Compute the constants of structure of the Lie algebra of $GL(n)$ with respect to the basis $\{\mathcal{E}_{11}, \mathcal{E}_{12}, \ldots, \mathcal{E}_{nn}\}$, where $\mathcal{E}_{ij}$ is the matrix with zeros in every place except the ith row and the jth column where there is a 1.

2. Choose bases for the Lie algebras of the following three Lie groups and compute the constants of structure of each Lie algebra with respect to its basis:

a. All $n \times n$ matrices of the form

$$\begin{pmatrix} 1 & x_{12} & \cdots & x_{1n} \\ 0 & 1 & \cdots & x_{2n} \\ \cdot & \cdot & \cdots & \cdot \\ 0 & 0 & \cdots & 1 \end{pmatrix}$$

b. All $n \times n$ matrices of the form

$$\begin{pmatrix} x_{11} & x_{12} & \cdots & x_{1n} \\ 0 & x_{22} & \cdots & x_{2n} \\ \cdots & \cdots & \cdots & \cdots \\ 0 & 0 & \cdots & x_{nn} \end{pmatrix}$$

c. The orthogonal group $O^+(n)$ on V^n.

3. Let $\mathfrak{A}$ be a 3-dimensional vector space with the basis $\{\alpha_1, \alpha_2, \alpha_3\}$. Show that $\mathfrak{A}$ may be given the structure of a Lie algebra in such a way that $[\alpha_1, \alpha_2] = \alpha_3$, $[\alpha_1, \alpha_3] = [\alpha_2, \alpha_3] = 0$. Show that every 2-dimensional subalgebra of $\mathfrak{A}$ must contain α_3.

CHAPTER 9

Fiber Bundles

This chapter is concerned with a generalization of the familiar concept of the product of manifolds. In order to understand more clearly how this generalization comes about, it is desirable to examine the product $M \times M'$ of two differentiable manifolds M and M' from the structural point of view. We shall find the answer to the question: Given a differentiable manifold N, how can we tell when it is the product manifold $M \times M'$?

9-1 PRODUCT MANIFOLDS

Let M and M' be differentiable manifolds, and let N be the product manifold $M \times M'$. The natural projections $\pi: N \to M$ and $\pi': N \to M'$ are differentiable mappings.

Let $m \in M$. The restriction of π' to the set $\pi^{-1}(m)$ is a one-to-one mapping of $\pi^{-1}(m)$ onto M'. Denote its *inverse* by ν'_m. Then $\nu'_m: M' \to N$ is a diffeomorphism which we call a natural injection of M' into N. There are also the analogous natural injections $\nu_{m'}: M \to N$ for each $m' \in M'$.

It is the existence of suitably interrelated projections $\pi: N \to M$ and $\pi': N \to M'$ which will tell us that N has the structure of $M \times M'$. The conditions are:

1. There are differentiable mappings $\pi: N \to M$ and $\pi': N \to M'$.
2. For each $m \in M$ the restriction of π' to $\pi^{-1}(m)$ is one-to-one and onto M'. Denote the inverse mapping by ν'_m.
3. For each $m' \in M'$ the restriction of π to $\pi'^{-1}(m')$ is one-to-one and onto M. Denote the inverse mapping by $\nu_{m'}$.
4. Each of the mappings $\nu'_m: M' \to N$ and each of the mappings $\nu_{m'}: M \to N$ is a diffeomorphism.

The reader may easily verify that under conditions 1 to 4 the mapping $n \to (\pi(n), \pi'(n))$ is a diffeomorphism of N onto $M \times M'$.

The notion of fiber bundle arises through a relaxation of conditions

1 to 4. In the first place, we shall give up the symmetry of the roles played by the manifolds M and M'. When N is a fiber bundle, there will be a projection $\pi: N \to M$ and injections $\nu'_m: M' \to N$ for each $m \in M$ but not generally the corresponding mappings when M and M' are interchanged.

EXERCISE

Show that condition 3 is superfluous.

9-2 COORDINATE BUNDLES

The notation of the previous section will now be modified. In place of N, M, and M' we shall write B, X, and F, respectively.

Let X, F, and B be three differentiable manifolds. Let π be a differentiable mapping of B onto X. π is called the *projection* of B onto the *base manifold* X. To produce a fiber bundle, we shall cover B suitably with a patchwork of product manifolds $U \times F$, where U is an open subset of X. This patchwork must be pieced together in a consistent manner, giving rise to certain compatibility conditions. Furthermore, we shall require that the pieces fit together smoothly and so as to exclude unmanageable pathological situations. We shall now describe these requirements more precisely.

Let X be covered by a family $\{U_\alpha\}$ of open subsets. For each set U_α we assume that there is a diffeomorphism η_α of the product manifold $U_\alpha \times F$ into B such that $\pi\circ\eta_\alpha(x, y) = x$ for each $x \in U_\alpha$ and $y \in F$ and that η_α maps $U_\alpha \times F$ onto $\pi^{-1}(U_\alpha)$. The manifold F is called the *fiber*.

Note that, if η_α is restricted to $\{x\} \times F$ with $x \in U_\alpha$, we obtain the analogue of the injection mapping of Section 9-1. However, this mapping generally depends also upon the index α. If it did not, we would eventually find, after all the conditions which we are about to describe are brought to bear, that we were led merely to the product manifold $X \times F$.

If $U_\alpha \cap U_\beta$ is not empty, the restriction $h_{\alpha\beta}$ of $\eta_\beta^{-1}\circ\eta_\alpha$ to $(U_\alpha \cap U_\beta) \times F$ is a diffeomorphism of this manifold. Then for fixed $x \in U_\alpha \cap U_\beta$ the mapping $(x, y) \to h_{\alpha\beta}(x, y)$ is a diffeomorphism of $\{x\} \times F$ and therefore gives rise naturally to a diffeomorphism of F, which we denote by $g_{\alpha\beta}(x)$. Thus we have the relation

$$h_{\alpha\beta}(x, y) = (x, g_{\alpha\beta}(x)y)$$

where we have written $g_{\alpha\beta}(x)y$ instead of $(g_{\alpha\beta}(x))(y)$. It is clear that if $U_\alpha \cap U_\beta \cap U_\gamma$ is not empty then

$$h_{\alpha\gamma} = h_{\beta\gamma} \circ h_{\alpha\beta}$$

on $(U_\alpha \cap U_\beta \cap U_\gamma) \times F$. It follows that

$$g_{\alpha\gamma}(x) = g_{\beta\gamma}(x) \circ g_{\alpha\beta}(x)$$

for each x in $U_\alpha \cap U_\beta \cap U_\gamma$.

The conditions that we have imposed so far will appear quite natural to the reader, who is accustomed to the similar requirements for defining a differentiable manifold. They are, however, not strong enough. In order to describe the desired smoothness requirements, we must introduce the notion of the action of a Lie group on a differentiable manifold.

Let $\mathfrak{G}$ be a Lie group, and let M be a differentiable manifold. We shall say that $\mathfrak{G}$ *acts on M on the left* or that $\mathfrak{G}$ is a *differentiable left transformation group* on M if the following conditions are satisfied:

1. With each $g \in \mathfrak{G}$ there is associated a diffeomorphism of M which we also denote by g. It must be borne in mind, however, that although g_1 and g_2 are different elements of $\mathfrak{G}$ they may be the same diffeomorphism of M.
2. The mapping of $\mathfrak{G} \times M$ into M defined by $(g, m) \to g(m)$ is differentiable.
3. The identity element of $\mathfrak{G}$ is the identity diffeomorphism of M.
4. $(g_1g_2)(m) = g_1(g_2(m))$ for all $g_1, g_2 \in \mathfrak{G}$ and $m \in M$.

We shall write gm for $g(m)$. Then condition 4 becomes

$$(g_1g_2)m = g_1(g_2m)$$

Action of $\mathfrak{G}$ on the right is defined by replacing condition 4 by $(g_1g_2)(m) = g_2(g_1(m))$. In this case we write mg for $g(m)$, and condition 4 becomes $m(g_1g_2) = (mg_1)g_2$.

$\mathfrak{G}$ will be said to act *effectively* on M if the only element of $\mathfrak{G}$ which is associated with the identity diffeomorphism of M is the identity element of $\mathfrak{G}$. In this case, distinct elements of $\mathfrak{G}$ are distinct diffeomorphisms of M.

In these terms it is possible to state the smoothness conditions that we desire. In the first place, we require that the mappings $g_{\alpha\beta}(x)$ be elements of a Lie group $\mathfrak{G}$ acting on the fiber F on the left. In addition, we require that the mapping $g_{\alpha\beta}$: $U_\alpha \cap U_\beta \to \mathfrak{G}$ given by $x \to g_{\alpha\beta}(x)$ be differentiable.

We shall summarize in the form of a definition the requirements which we have been discussing.

A *coordinate bundle* consists of the following differentiable manifolds and mappings:

1. Three differentiable manifolds B, X, and F.
2. A differentiable mapping π of B onto X.
3. A Lie group $\mathfrak{G}$ which acts effectively on F on the left.
4. An indexed family $\{U_\alpha\}$, $\alpha \in A$, of open sets which covers X.
5. For each α in A, a diffeomorphism η_α of the product manifold $U_\alpha \times F$ onto $\pi^{-1}(U_\alpha)$, subject to the following conditions:
6. $(\pi \circ \eta_\alpha)(x, y) = x$ for $(x, y) \in U_\alpha \times F$.
7. For $x \in U_\alpha \cap U_\beta$, the diffeomorphism $y \to \theta[(\eta_\beta^{-1} \circ \eta_\alpha)(x, y)]$, where θ: $U_\beta \times F \to F$ is the natural projection, is given by an element $g_{\alpha\beta}(x)$ of $\mathfrak{G}$.
8. For each pair $\alpha, \beta \in A$ such that $U_\alpha \cap U_\beta$ is not empty, the mapping $x \to g_{\alpha\beta}(x)$ of $U_\alpha \cap U_\beta$ into $\mathfrak{G}$ is differentiable.

B, X, and F are called the bundle manifold, base manifold, and fiber, respectively. π is called the projection. The mappings $g_{\alpha\beta}$ are called the coordinate transformations of the coordinate bundle.

As an example, the reader may consider the tangent bundle $\mathfrak{T}(M)$ of a differentiable manifold M. In this case $\mathfrak{T}(M)$ is the bundle manifold and M is the base manifold. If M has dimension n, the fiber is V^n. The Lie group which acts on V^n is the general linear group $GL(n)$. For the indexed family $\{U_\alpha\}$ we may take any covering of M by coordinate neighborhoods. Let U_α be such a coordinate neighborhood and $x_1, \ldots, x_n$ the coordinates in U_α. Then the diffeomorphism η_α is given by

$$\eta_\alpha(x_1, \ldots, x_n, a_1, \ldots, a_n) = \left(x_1, \ldots, x_n, a_1 \frac{\partial}{\partial x_1} + \cdots + a_n \frac{\partial}{\partial x_n}\right)$$

The reader may easily verify that the conditions of the definition are satisfied.

9-3 THE CONSTRUCTION OF COORDINATE BUNDLES

The definition of a coordinate bundle has been given from the structural point of view. But it is natural to ask how we may start with two differentiable manifolds X and F and construct a coordinate bundle with these as base manifold and fiber, respectively. Of course we shall need a Lie group $\mathfrak{G}$ which acts on F. It turns out that it is sufficient to have a suitable open covering $\{U_\alpha\}$ of X and suitable differentiable mappings $g_{\alpha\beta}$: $U_\alpha \cap U_\beta \to \mathfrak{G}$. Then we may construct the bundle manifold B, the projection π, and the mappings $\eta_\alpha\colon U_\alpha \times F \to B$ so as to satisfy the definition of a coordinate bundle.

We shall assume that we are given the following objects:

1. Differentiable manifolds X and F
2. A Lie group $\mathfrak{G}$ which acts effectively on F on the left
3. An open covering $\{U_\alpha\}$ of X
4. For each pair α, β of indices such that $U_\alpha \cap U_\beta$ is not empty, a differentiable mapping $g_{\alpha\beta}\colon U_\alpha \cap U_\beta \to \mathfrak{G}$

We assume that these mappings satisfy the compatibility conditions:

5. $g_{\beta\gamma}(x)g_{\alpha\beta}(x) = g_{\alpha\gamma}(x)$ for $x \in U_\alpha \cap U_\beta \cap U_\gamma$

It follows that $g_{\alpha\alpha}(x)$ is the identity element of $\mathfrak{G}$ and that $g_{\beta\alpha}(x)$ and $g_{\alpha\beta}(x)$ are inverses of one another.

The family $\{g_{\alpha\beta}\}$ will be called a *system of coordinate transformations on X with values in $\mathfrak{G}$*. The use of the word "coordinate" in the language of fiber bundles is by analogy with its use in the language of differentiable manifolds. We do not have coordinates here in the strict sense of the word.

Let A be the index set of the covering $\{U_\alpha\}$. Give the set A the discrete topology (in which every set is open), and form the topological product $X \times F \times A$. Let H be the subset of $X \times F \times A$ consisting of those points (x, y, α) such that $x \in U_\alpha$. Thus H is the union of the subsets $U_\alpha \times F \times \{\alpha\}$ of $X \times F \times A$ as α runs through A. Each of these is an open subset of H.

In H define $(x, y, \alpha) \sim (x', y', \beta)$ when $x = x'$ and $g_{\alpha\beta}(x)y = y'$. From condition 5 it follows that this is an equivalence relation. Let B be the set of equivalence classes. Let $\theta\colon H \to B$ be the natural mapping which assigns to $h \in H$ the equivalence class which contains h. Then θ is one-to-one on each of the subsets $U_\alpha \times F \times \{\alpha\}$.

Give to B the topology which is naturally associated with θ, in which a subset W of B is open when $\theta^{-1}(W)$ is open in H. Then $\theta\colon H \to B$ is continuous, and the reader may show that it is an open mapping. Consequently its restriction to each of the open subsets $U_\alpha \times F \times \{\alpha\}$ is a homeomorphism.

Define $\pi_1\colon H \to X$ by $\pi_1(x, y, \alpha) = x$. Then π_1 induces a mapping $\pi\colon B \to X$ such that $\pi \circ \theta = \pi_1$ (see Figure 9-1). Since π_1 is continuous and θ is an open mapping, it follows that π is continuous. Furthermore, since $\pi_1{}^{-1}(U_\alpha) = U_\alpha \times F \times \{\alpha\}$, it follows that

$$\pi^{-1}(U_\alpha) = \theta(U_\alpha \times F \times \{\alpha\})$$

Define the mapping $\eta_\alpha\colon U_\alpha \times F \to \pi^{-1}(U_\alpha)$ by

$$\eta_\alpha(x, y) = \theta(x, y, \alpha)$$

Then η_α maps $U_\alpha \times F$ onto $\pi^{-1}(U_\alpha)$. Since η_α is the composition of the natural homeomorphism of $U_\alpha \times F$ onto $U_\alpha \times F \times \{\alpha\}$ and the restriction of θ to $U_\alpha \times F \times \{\alpha\}$ (see Figure 9-2), it follows that η_α is a homeomorphism of $U_\alpha \times F$ onto $\pi^{-1}(U_\alpha)$.

The topological space B is Hausdorff. Since the covering $\{U_\alpha\}$ of X has a denumerable refinement, it follows that B also has a denumerable basis. We wish to give B the structure of a differentiable manifold in

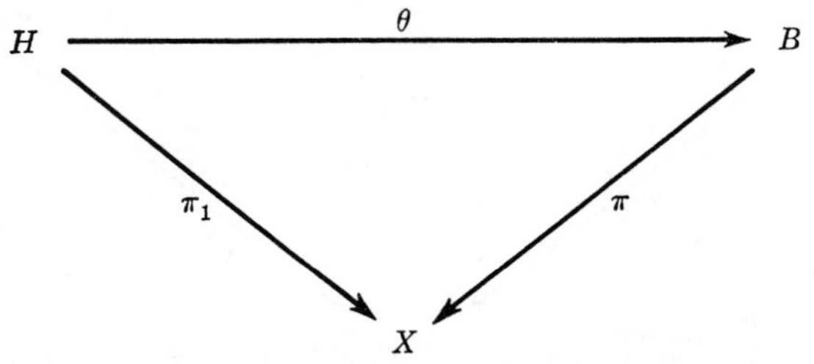

FIG. 9-1

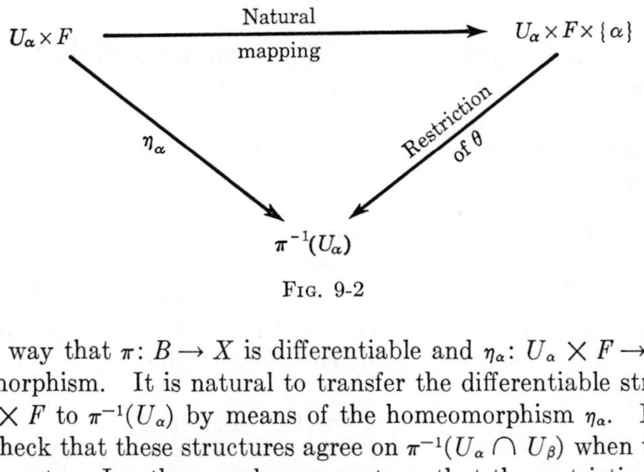

FIG. 9-2

such a way that $\pi: B \to X$ is differentiable and $\eta_\alpha: U_\alpha \times F \to B$ is a diffeomorphism. It is natural to transfer the differentiable structure of $U_\alpha \times F$ to $\pi^{-1}(U_\alpha)$ by means of the homeomorphism η_α. But we must check that these structures agree on $\pi^{-1}(U_\alpha \cap U_\beta)$ when this set is not empty. In other words, we must see that the restriction $h_{\alpha\beta}$ of $\eta_\beta^{-1} \circ \eta_\alpha$ to $(U_\alpha \cap U_\beta) \times F$ is a diffeomorphism.

The mapping $h_{\alpha\beta}$ is given by

$$h_{\alpha\beta}(x, y) = (x, g_{\alpha\beta}(x)y)$$

(see Figure 9-3). It suffices to show that the mappings $x \to h_{\alpha\beta}(x, y)$ for fixed y and $y \to h_{\alpha\beta}(x, y)$ for fixed x are differentiable. For these statements imply that $h_{\alpha\beta}$ is differentiable. Since $h_{\beta\alpha}$ is the inverse of $h_{\alpha\beta}$, it follows that $h_{\alpha\beta}$ is a diffeomorphism.

From condition 4, $x \to g_{\alpha\beta}(x)$ is differentiable, and from condition 2, $g \to gy$ is differentiable. The composition of these is the mapping $x \to g_{\alpha\beta}(x)y$, which is therefore differentiable. It follows that $x \to (x, g_{\alpha\beta}(x)y)$ is differentiable. A similar, but simpler, argument shows that $y \to (x, g_{\alpha\beta}(x)y)$ is differentiable.

The remainder of conditions 1 to 8 for a coordinate bundle are easily verified. The results may be summarized in the form of a theorem.

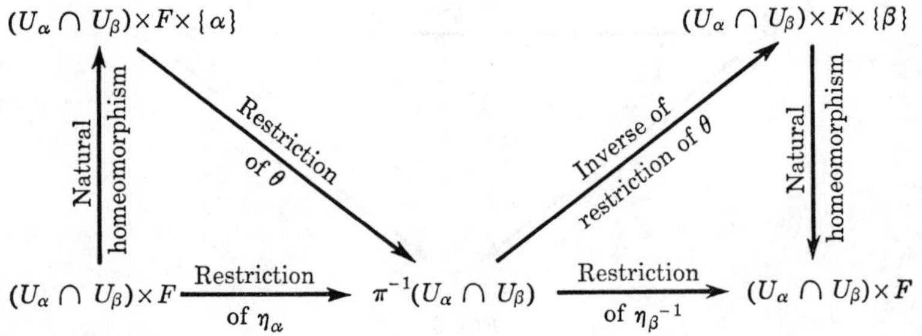

FIG. 9-3

Theorem 9-1. *If X and F are differentiable manifolds, $\mathfrak{G}$ is a Lie group which acts effectively on F on the left, and $\{U_\alpha, g_{\alpha\beta}\}$ is a system of coordinate transformations on X with values in $\mathfrak{G}$, then there is a coordinate bundle with base manifold X, fiber F, group $\mathfrak{G}$, and coordinate transformations $\{U_\alpha, g_{\alpha\beta}\}$.*

The coordinate bundle provided by this theorem is essentially unique. Suppose that B and B' are two bundle manifolds with projections $\pi: B \to X$ and $\pi': B' \to X$ and mappings $\eta_\alpha: U_\alpha \times F \to B$ and $\eta'_\alpha: U_\alpha \times F \to B'$, respectively, which satisfy the conclusion of Theorem 9-1. Then we may define diffeomorphisms

$$\sigma_\alpha = \eta'_\alpha \circ \eta_\alpha^{-1}: \pi^{-1}(U_\alpha) \to \pi'^{-1}(U_\alpha)$$

The reader may easily show that for $z \in \pi^{-1}(U_\alpha \cap U_\beta)$ we have $\sigma_\alpha(z) = \sigma_\beta(z)$. Hence the diffeomorphisms $\{\sigma_\alpha\}$ determine a mapping $\sigma: B \to B'$ such that $\sigma(z) = \sigma_\alpha(z)$ for $z \in \pi^{-1}(U_\alpha)$. σ is a diffeomorphism of B onto B' which satisfies $\pi' \circ \sigma = \pi$. For this reason it is pointless to distinguish between these two coordinate bundles, and we shall denote either one of them by $\{B, X, F, \mathfrak{G}, \{U_\alpha, g_{\alpha\beta}\}\}$.

Observe that in Theorem 9-1 the system of coordinate transformations $\{U_\alpha, g_{\alpha\beta}\}$ refers only to the Lie group $\mathfrak{G}$. Consequently, with the same system of coordinate transformations on X, it is possible to construct different coordinate bundles by altering the fiber F and the action of $\mathfrak{G}$ on F. In particular, if $\mathfrak{B} = \{B, X, F, \mathfrak{G}, \{U_\alpha, g_{\alpha\beta}\}\}$ is a given coordinate bundle, we may obtain a new coordinate bundle by replacing F by $\mathfrak{G}$ itself and taking the action of $\mathfrak{G}$ on $\mathfrak{G}$ to be multiplication on the left by elements of $\mathfrak{G}$. The resulting coordinate bundle $\mathfrak{P} = \{P, X, \mathfrak{G}, \mathfrak{G}, \{U_\alpha, g_{\alpha\beta}\}\}$ is called the *principal coordinate bundle* associated with $\mathfrak{B}$.

EXERCISES

1. Complete the proof of Theorem 9-1 and the uniqueness statement which follows it.

2. Verify that the principal bundle of a differentiable manifold M is the principal coordinate bundle associated with the tangent bundle $\mathfrak{T}(M)$.

3. Let $\mathfrak{B} = \{B, X, F, \mathfrak{G}, \{U_\alpha, g_{\alpha\beta}\}\}$ be a coordinate bundle. Let $\psi: \mathfrak{G} \to \mathfrak{H}$ be a Lie homomorphism into a Lie group $\mathfrak{H}$. Show that ψ and $\mathfrak{B}$ determine a principal coordinate bundle with group $\mathfrak{H}$ and base manifold X.

9-4 FIBER BUNDLES

The alert reader, guided by the analogy with differentiable manifolds, will expect that it is possible to eliminate the arbitrariness in the choice of the covering $\{U_\alpha\}$ of the base manifold X and the mappings $\eta_\alpha: U_\alpha \times F \to B$. Indeed, this can be done, as it was for differentiable manifolds, by introducing *all* coordinate neighborhoods and mappings which are compatible with the given ones. However, we shall describe another method which is quite natural and which entirely circumvents the covering $\{U_\alpha\}$.

It suffices to devise such a method for principal coordinate bundles. For if we are given a principal coordinate bundle $\mathfrak{P}$ with group $\mathfrak{G}$, then every coordinate bundle associated with $\mathfrak{P}$ arises by replacing the fiber $\mathfrak{G}$ by another manifold F on which $\mathfrak{G}$ acts effectively on the left. We shall see later how the bundle manifold of such a coordinate bundle may be described directly in terms of $\mathfrak{P}$, F, and $\mathfrak{G}$.

Let $\mathfrak{P} = \{P, X, \mathfrak{G}, \mathfrak{G}, \{U_\alpha, g_{\alpha\beta}\}\}$ be a principal coordinate bundle. We shall see that $\mathfrak{G}$ acts on P in a natural way on the right. In the first place, $\mathfrak{G}$ acts on the manifold $U_\alpha \times \mathfrak{G}$ on the right in accordance with the rule

$$(x, g_1)g_2 = (x, g_1g_2)$$

Let $\eta_\alpha: U_\alpha \times \mathfrak{G} \to P$ be the coordinate mapping corresponding to U_α. We would like to define the action of $\mathfrak{G}$ on $\eta_\alpha(x, g_1)$ by

$$[\eta_\alpha(x, g_1)]g_2 = \eta_\alpha(x, g_1g_2)$$

This will be well-defined if $\eta_\beta(x, g_1') = \eta_\alpha(x, g_1)$ implies

$$\eta_\beta(x, g_1'g_2) = \eta_\alpha(x, g_1g_2)$$

The reader may easily verify this and also the following simple properties of this action:

1. If $\pi: P \to X$ is the projection and $g \in \mathfrak{G}$, then g maps the set $\pi^{-1}(x)$ onto itself.

2. If g is not the identity element of $\mathfrak{G}$, then g leaves no element of P fixed.

3. If $z_1, z_2 \in \pi^{-1}(x)$, there is an element g of $\mathfrak{G}$ such that $z_2 = z_1g$.

If M is any differentiable manifold on which $\mathfrak{G}$ acts as a differentiable right transformation group, we may define an equivalence relation in M as follows: Let $z_1, z_2 \in M$, and define $z_1 \sim z_2$ if there is a g in $\mathfrak{G}$ such that $z_2 = z_1g$. The set of equivalence classes in M will be denoted by $M/\mathfrak{G}$.

In the case of the principal coordinate bundle $\{P, X, \mathfrak{G}, \mathfrak{G}, \{U_\alpha, g_{\alpha\beta}\}\}$ we may form $P/\mathfrak{G}$. Let $\pi_1: P \to P/\mathfrak{G}$ be the natural mapping which carries $z \in P$ onto the equivalence class which contains z. Properties 1 and 3 assure us that there is a one-to-one mapping $\tau: X \to P/\mathfrak{G}$ such that $\pi_1 \circ \tau = \pi$. We may use the mapping τ to transfer the differentiable structure of X to $P/\mathfrak{G}$. Then the manifold $P/\mathfrak{G}$ may be used in place of X as the base manifold of P if we replace the projection $\pi: P \to X$ by the new projection $\pi_1: P \to P/\mathfrak{G}$.

The main problem is therefore to characterize those actions of a Lie group $\mathfrak{G}$ on a manifold P on the right for which the natural mapping $\pi_1: P \to P/\mathfrak{G}$ may be considered as the projection mapping of a suitable principal bundle. The central idea which we require is that of a local cross section.

Let B and X be differentiable manifolds, and let π be a differentiable mapping of B onto X. Let U be an open subset of X. A differentiable mapping $\sigma: U \to B$ such that $\pi \circ \sigma =$ identity on U will be called a *local cross section* of B. If $U = X$, then σ is called simply a cross section. As examples, consider the differentiable vector fields on open subsets of a differentiable manifold M. These are local cross sections of the tangent bundle $\mathfrak{T}(M)$.

Let $\{B, X, F, \mathfrak{G}, \{U_\alpha, g_{\alpha\beta}\}\}$ be a coordinate bundle. Let $\eta_\alpha: U_\alpha \times F \to B$ be the mapping associated with U_α. Then the composition of the injection $x \to (x, y)$ of U_α onto $U_\alpha \times \{y\}$ and the mapping η_α provides us with a local cross section of B over U_α. We see, therefore, that *in a coordinate bundle every point of the base manifold has a neighborhood over which there is a local cross section.*

Let P be a differentiable manifold, and let $\mathfrak{G}$ be a Lie group which acts on P on the right. Let $P/\mathfrak{G}$ have a differentiable structure such that the natural mapping $\pi: P \to P/\mathfrak{G}$ is differentiable. If every point x in $P/\mathfrak{G}$ has a neighborhood U over which there is a local cross section $\sigma: U \to P$ such that the mapping $(x, g) \to \sigma(x)g$ is a diffeomorphism of $U \times \mathfrak{G}$ into P, we shall say that π is *locally trivial.*

Let us return to the principal bundle $\{P, P/\mathfrak{G}, \mathfrak{G}, \mathfrak{G}, \{U_\alpha, g_{\alpha\beta}\}\}$. Let $g_0 \in \mathfrak{G}$. Then $\sigma: U_\alpha \to P$ defined by $\sigma(x) = \eta_\alpha(x, g_0)$ is a local cross section, and the mapping $(x, g) \to \sigma(x)g$ is the same as $(x, g) \to \eta_\alpha(x, g_0g)$. It follows that the projection $\pi: P \to P/\mathfrak{G}$ is locally trivial. The following theorem shows that this condition of local triviality is a satisfactory substitute for the former smoothness conditions which we had imposed on a principal coordinate bundle and it does away with the coordinate mappings.

Theorem 9-2. *Let P be a differentiable manifold, and let $\mathfrak{G}$ be a Lie group which acts on P on the right with the following results:*

(i) *If g is not the identity element of $\mathfrak{G}$, then g leaves no element of P fixed.*

(ii) *$P/\mathfrak{G}$ has a differentiable structure such that the natural projection $\pi: P \to P/\mathfrak{G}$ is differentiable and locally trivial.*

Then there is a principal coordinate bundle with bundle manifold P, base manifold $P/\mathfrak{G}$, projection π, and group $\mathfrak{G}$ and such that the action of $\mathfrak{G}$ on P in this coordinate bundle is the given action. Furthermore, the differentiable structure of $P/\mathfrak{G}$ is uniquely determined.

Proof. If $x \in P/\mathfrak{G}$, there are an open set U_x in $P/\mathfrak{G}$ and a local cross section $\sigma_x: U_x \to P$ such that $(u, g) \to \sigma_x(u)g$ is a diffeomorphism of $U_x \times \mathfrak{G}$ into P. We shall take the family $\{U_x\}$, indexed by the points of $P/\mathfrak{G}$, as the covering of $P/\mathfrak{G}$. Define $\eta_x: U_x \times \mathfrak{G} \to P$ by

$$\eta_x(u, g) = \sigma_x(u)g$$

η_x is a diffeomorphism of $U_x \times \mathfrak{G}$ onto $\pi^{-1}(U_x)$.

Let $u \in U_{x_1} \cap U_{x_2}$. If $g \in \mathfrak{G}$, there is a $g' \in \mathfrak{G}$ such that $\sigma_{x_2}(u)g' = \sigma_{x_1}(u)g$. Let $g_{x_1x_2}(u) = g'g^{-1}$. Then $\sigma_{x_2}(u)g_{x_1x_2}(u) = \sigma_{x_1}(u)$, and $g_{x_1x_2}(u)$ is uniquely determined by u. The three mappings $\sigma_{x_1}: U_{x_1} \cap U_{x_2} \to P$, $\eta_{x_2}^{-1}: \pi^{-1}(U_{x_1} \cap U_{x_2}) \to (U_{x_1} \cap U_{x_2}) \times \mathfrak{G}$, and the natural mapping $(U_{x_1} \cap U_{x_2}) \times \mathfrak{G} \to \mathfrak{G}$ are all differentiable; hence their composition $g_{x_1x_2}: U_{x_1} \cap U_{x_2} \to \mathfrak{G}$ is differentiable. Furthermore, $\eta_{x_1}(u, g) = \eta_{x_2}(u, g_{x_1x_2}(u)g)$. All the conditions 1 to 8 of the definition of a coordinate bundle are easily verified.

Assume that it is possible to put two differentiable structures on $P/\mathfrak{G}$ in such a way that the hypotheses of the theorem are satisfied. Denote the two resulting manifolds by X and X' and the natural mapping of P onto $P/\mathfrak{G}$ by $\pi: P \to X$ and $\pi': P \to X'$, respectively, in the two cases. Then there is a one-to-one mapping ψ of X onto X' such that $\pi' = \psi \circ \pi$. As a mapping of sets, ψ is the identity mapping of $P/\mathfrak{G}$. We shall show that $\psi: X \to X'$ is a diffeomorphism. Let $x \in X$. There are an open set U containing x and a local cross section $\sigma: U \to P$. Then the restriction of ψ to U is given by $\pi' \circ \sigma$, which is differentiable at x. Similarly ψ^{-1} is also differentiable, and Theorem 9-2 is proved.

Let $\mathfrak{P} = \{P, X, \mathfrak{G}, \mathfrak{G}, \{U_\alpha, g_{\alpha\beta}\}\}$ and $\mathfrak{P}' = \{P, X', \mathfrak{G}, \mathfrak{G}, \{U'_{\alpha'}, g'_{\alpha'\beta'}\}\}$ be two principal coordinate bundles with the same bundle manifold and the same group. We shall say that $\mathfrak{P}$ and $\mathfrak{P}'$ are *equivalent* if they give rise to the same action of $\mathfrak{G}$ on P on the right. An equivalence class of principal coordinate bundles will be called a *principal bundle.*

Theorem 9-3. *$\{P, X, \mathfrak{G}, \mathfrak{G}, \{U_\alpha, g_{\alpha\beta}\}\}$ is equivalent to $\{P, X, \mathfrak{G}, \mathfrak{G}, \{U_\alpha, g'_{\alpha\beta}\}\}$ if and only if there are differentiable mappings $\lambda_\alpha: U_\alpha \to \mathfrak{G}$ such that*

$$g'_{\alpha\beta}(x) = [\lambda_\beta(x)]^{-1} g_{\alpha\beta}(x) \lambda_\alpha(x)$$

for $x \in U_\alpha \cap U_\beta$.

Proof. Let $\eta_\alpha: U_\alpha \times \mathfrak{G} \to P$ and $\eta'_\alpha: U_\alpha \times \mathfrak{G} \to P$ be the coordinate mappings corresponding to the coordinate transformations $\{g_{\alpha\beta}\}$ and $\{g'_{\alpha\beta}\}$, respectively.

Assume that the given coordinate bundles are equivalent. If $x \in U_\alpha$, the element $\lambda_\alpha(x)$ of $\mathfrak{G}$ is well-defined by the relation

$$\eta'_\alpha(x, e) = \eta_\alpha(x, \lambda_\alpha(x))$$

where e is the identity element of $\mathfrak{G}$. Since $\eta_\alpha^{-1} \circ \eta'_\alpha$ is a diffeomorphism of $U_\alpha \times \mathfrak{G}$, $\lambda_\alpha: U_\alpha \to \mathfrak{G}$ is differentiable. The equivalence gives

$$\eta'_\alpha(x,g) = \eta_\alpha(x, \lambda_\alpha(x) g)$$

for $g \in \mathfrak{G}$ and hence the desired relation

$$g'_{\alpha\beta}(x) = [\lambda_\beta(x)]^{-1} g_{\alpha\beta}(x) \lambda_\alpha(x)$$

for $x \in U_\alpha \cap U_\beta$.

Conversely, if there are differentiable mappings $\lambda_\alpha: U_\alpha \to \mathfrak{G}$ such that

$$g'_{\alpha\beta}(x) = [\lambda_\beta(x)]^{-1} g_{\alpha\beta}(x) \lambda_\alpha(x)$$

for $x \in U_\alpha \cap U_\beta$, consider the mapping $\mu_\alpha: U_\alpha \times \mathfrak{G} \to P$ defined by

$$\mu_\alpha(x, g) = \eta_\alpha(x, \lambda_\alpha(x) g)$$

μ_α is one-to-one and differentiable. Its inverse is given by the composition of the mappings η_α^{-1} and $(x, g) \rightarrow (x, [\lambda_\alpha(x)]^{-1}g)$ and hence is also differentiable. Thus $\mu_\alpha\colon U_\alpha \times \mathfrak{G} \rightarrow P$ is a diffeomorphism. Furthermore,

$$\mu_\alpha(x, g) = \mu_\beta(x, g'_{\alpha\beta}(x)g)$$

and so the coordinate mappings $\mu_\alpha\colon U_\alpha \times \mathfrak{G} \rightarrow P$ define the bundle $\mathfrak{P}'$. It is clear that the action of $\mathfrak{G}$ on P is the same for $\mathfrak{P}'$ as it is for $\mathfrak{P}$.

Theorem 9-3 actually serves to test the equivalence of any two principal coordinate bundles. For equivalent principal coordinate bundles may be regarded as having the same base manifold. If at first they do not have the same coordinate coverings of this base manifold, these coverings may be replaced by a common refinement, whereupon we may apply Theorem 9-3.

Two coordinate bundles $\{B, X, F, \mathfrak{G}, \{U_\alpha, g_{\alpha\beta}\}\}$ and $\{B', X', F, \mathfrak{G}, \{U'_{\alpha'}, g'_{\alpha'\beta'}\}\}$ with the same fiber and group and action of the group on the fiber will be called *equivalent* if they have equivalent principal coordinate bundles. An equivalence class of coordinate bundles will be called a *fiber bundle*. Of course, equivalent coordinate bundles may be regarded as having the same base manifold, and we shall now see that they may also be regarded as having the same bundle manifold.

Let $\mathfrak{B} = \{B, X, F, \mathfrak{G}, \{U_\alpha, g_{\alpha\beta}\}\}$ be a coordinate bundle. Let $\mathfrak{P} = \{P, X, \mathfrak{G}, \mathfrak{G}, \{U_\alpha, g_{\alpha\beta}\}\}$ be the associated principal coordinate bundle. Then $\mathfrak{G}$ acts on F on the left and on P on the right.

Introduce the product manifold $P \times F$. We wish to map $P \times F$ into B. Let $\eta_\alpha\colon U_\alpha \times F \rightarrow B$ and $\mu_\alpha\colon U_\alpha \times \mathfrak{G} \rightarrow P$ be the coordinate mappings of $\mathfrak{B}$ and $\mathfrak{P}$, respectively. If $(z, y) \in P \times F$, we may suppose that $z = \mu_\alpha(x, g)$. We wish to define $\theta(z, y) = \eta_\alpha(x, gy)$, and so we must show that this definition is independent of α. If $z = \mu_\beta(x, g')$, then $g' = g_{\alpha\beta}(x)g$ and $\eta_\beta(x, g'y) = \eta_\beta(x, g_{\alpha\beta}gy) = \eta_\alpha(x, gy)$. Thus $\theta\colon P \times F \rightarrow B$ is well-defined. Clearly θ maps $P \times F$ onto B.

Let $\theta(z_1, y_1) = \theta(z_2, y_2)$. We may suppose that $\theta(z_1, y_1) = \eta_\alpha(x, g_1y_1)$ and $\theta(z_2, y_2) = \eta_\alpha(x, g_2y_2)$, where $z_1 = \mu_\alpha(x, g_1)$ and $z_2 = \mu_\alpha(x, g_2)$. Let $g = g_1^{-1}g_2$. It follows that $(z_2, y_2) = (z_1g, g^{-1}y_1)$. Conversely, if $(z_2, y_2) = (z_1g, g^{-1}y_1)$ for some element g from $\mathfrak{G}$, then $\theta(z_1, y_1) = \theta(z_2, y_2)$. If we define the action of $\mathfrak{G}$ on $P \times F$ by

$$(z, y)g = (zg, g^{-1}y)$$

then $\mathfrak{G}$ acts on the right. We have just seen that, if $\nu\colon P \times F \rightarrow (P \times F)/\mathfrak{G}$ is the natural mapping, there is a one-to-one correspondence θ_1 of $(P \times F)/\mathfrak{G}$ onto B such that $\theta = \theta_1 \circ \nu$. If we transfer the differentiable structure of B to $(P \times F)/\mathfrak{G}$, we may use $(P \times F)/\mathfrak{G}$

as the bundle manifold of $\mathfrak{B}$, provided that we replace the projection $\pi: B \to X$ by $\pi \circ \theta_1$. If $\rho: P \to X$ is the projection of $\mathfrak{P}$, then

$$\pi \circ \theta(z, y) = \rho(z)$$

Finally we shall prove that $\theta: P \times F \to B$ is differentiable and locally trivial. It follows from this that $P \times F$ and $\mathfrak{G}$ define a principal bundle over the base manifold $(P \times F)/\mathfrak{G}$ and consequently the differentiable structure on $(P \times F)/\mathfrak{G}$ is unique.

The mapping $\theta: P \times F \to B$ is given locally by the composition of the mappings

$$\begin{aligned} (z, y) &\to (\mu_\alpha^{-1}(z), y) \\ ((x, g), y) &\to (x, g, y) \\ (x, g, y) &\to \eta_\alpha(x, gy) \end{aligned}$$

and is therefore differentiable. On the other hand, the composition of the mappings η_α^{-1} and $(x, y) \to (\mu_\alpha(x, e), y)$, where e is the identity element of $\mathfrak{G}$, provides us with a local cross section $\sigma: \pi^{-1}(U_\alpha) \to P \times F$. Consider the mapping $\tau: \pi^{-1}(U_\alpha) \times \mathfrak{G} \to P \times F$ given by $\tau(z, g) = \sigma(z)g$. Then

$$\tau(\eta_\alpha(x, y), g) = (\mu_\alpha(x, g), g^{-1}y)$$

for $x \in U_\alpha$, $y \in F$, $g \in \mathfrak{G}$. It is clear that τ is differentiable and one-to-one. The inverse mapping τ^{-1} is the composition of the mappings

$$\begin{aligned} (w, y) &\to (\mu_\alpha^{-1}(w), y) \\ ((x, g), y) &\to (x, g, gy) \\ (x, g, y) &\to (\eta_\alpha(x, y), g) \end{aligned}$$

and so is also differentiable. Thus θ is locally trivial.

9-5 HOMOGENEOUS SPACES

Let $\mathfrak{G}$ be a Lie group and $\mathfrak{H}$ a closed Lie subgroup of $\mathfrak{G}$. Actually we need only assume that $\mathfrak{H}$ is a closed subgroup. For it is possible to prove that a closed subgroup of a Lie group is also a Lie group, but we shall not present the proof.

Through group multiplication, $\mathfrak{H}$ acts on the manifold $\mathfrak{G}$ on the right, and so we may form the set $\mathfrak{G}/\mathfrak{H}$. $\mathfrak{G}/\mathfrak{H}$ consists of the left cosets of $\mathfrak{H}$ in $\mathfrak{G}$. The action of $\mathfrak{H}$ on $\mathfrak{G}$ is such that the only element of $\mathfrak{H}$ which leaves an element of $\mathfrak{G}$ fixed is the identity element. We shall show that the set $\mathfrak{G}/\mathfrak{H}$ may be given the structure of a differentiable manifold in such a way that the natural mapping $\pi: \mathfrak{G} \to \mathfrak{G}/\mathfrak{H}$ is differentiable and locally trivial. It then follows that $\mathfrak{G}$ is a principal bundle with the group $\mathfrak{H}$. With this differentiable structure, which we have seen is uniquely determined, $\mathfrak{G}/\mathfrak{H}$ is called a *homogeneous space.*

Lemma 9-1. *Let $\mathfrak{G}$ be a Lie group of dimension n, and let $\mathfrak{H}$ be a connected closed (Lie) subgroup of $\mathfrak{G}$ of dimension p. Then there is a coordinate neighborhood V of the identity element of $\mathfrak{G}$ with coordinates $(x_1, \ldots, x_n)$ such that $\mathfrak{H} \cap V$ is the set of points of V with $x_{p+1} = \cdots = x_n = 0$.*

Proof. Let e be the identity element of $\mathfrak{G}$. The tangent space to $\mathfrak{H}$ at e generates a left-invariant p-vector field X on $\mathfrak{G}$. Since X is integrable, there is a coordinate neighborhood U of e in $\mathfrak{G}$ with coordinates $(x_1, \ldots, x_n)$ such that $e = (0, \ldots, 0)$, U has the form

$$U = \{(x_1, \ldots, x_n) | \, x_1^2 + \cdots + x_n^2 < r^2\}$$

and X is spanned by $\partial/\partial x_1, \ldots, \partial/\partial x_p$ on U. $\mathfrak{H}$ is a maximal integral manifold for X.

Let N be the submanifold of U defined by the equations

$$x_1 = \cdots = x_p = 0$$

Then $e \in N$. Let $\theta: U \to N$ be defined by

$$\theta(x_1, \ldots, x_n) = (0, \ldots, 0, x_{p+1}, \ldots, x_n)$$

If $z \in N$, then $\theta^{-1}(z)$ is a connected integral manifold for X on U. Thus if $\theta^{-1}(z)$ contains a point of $\mathfrak{H}$, then $\theta^{-1}(z)$ is an open subset of $\mathfrak{H}$ and z itself is in $\mathfrak{H}$. Hence

$$\mathfrak{H} \cap U = \bigcup_{z \in \mathfrak{H} \cap N} \theta^{-1}(z)$$

and this is a union of disjoint sets. Since $\mathfrak{H}$ has a denumerable basis, it follows that $\mathfrak{H} \cap N$ is a denumerable set.

Consider $\mathfrak{H} \cap N = \{z_1, z_2, \ldots\}$ in the topology of $\mathfrak{G}$. We shall prove that $\mathfrak{H} \cap N$ must contain an isolated point. Suppose, on the contrary, that every point of $\mathfrak{H} \cap N$ is a point of accumulation of $\mathfrak{H} \cap N$. The reader may easily construct a sequence $U_1 \supset U_2 \supset \cdots$ of open sets such that $\bar{U}_1 \subset U$, U_j contains a point of $\mathfrak{H} \cap N$ but does not contain z_j, $U_{j+1} \subset U_j$, and, in terms of the natural metric on U, the diameter of $U_{j+1} \leqq$ one-half the diameter of U_j, for $j = 1, 2, \ldots$. Then let $\{w\} = \bigcap \bar{U}_j$. On the one hand, $w \notin \mathfrak{H} \cap N$. On the other hand, w is a point of accumulation of $\mathfrak{H} \cap N$. Since $\mathfrak{H}$ is closed, $w \in \mathfrak{H}$. Since $w \in \bar{U}_1 \subset U$ and $w \in \bar{N}$, $w \in N$. But this implies $w \in \mathfrak{H} \cap N$, which is a contradiction.

We may now prove that every point of $\mathfrak{H} \cap N$ is isolated. Let h_0

be an isolated point of $\mathfrak{H} \cap N$, and let $h \in \mathfrak{H} \cap N$. There is a neighborhood W of h_0 in $\mathfrak{H}$ such that $hh_0^{-1}W \subset \theta^{-1}(h)$ because $\theta^{-1}(h)$ is an open set in $\mathfrak{H}$. We may assume that $W \subset \theta^{-1}(h_0)$. But h_0 is isolated in $\mathfrak{H} \cap N$ in the topology on $\mathfrak{G}$. Hence there is an open set Z in $\mathfrak{G}$ such that $W = \mathfrak{H} \cap Z$. Then $hh_0^{-1}W = \mathfrak{H} \cap hh_0^{-1}Z$. Since $hh_0^{-1}Z$ is open in $\mathfrak{G}$, this shows that h is also an isolated point of $\mathfrak{H} \cap N$.

In particular, e is isolated in $\mathfrak{H} \cap N$, and there is an open set V in $\mathfrak{G}$, $V \subset U$, such that $\theta^{-1}(e) = \mathfrak{H} \cap V$, which proves the lemma.

Corollary. *The topology induced on $\mathfrak{H}$ as a subspace of $\mathfrak{G}$ and the topology of $\mathfrak{H}$ as a maximal integral manifold are the same.*

Let $\mathfrak{G}$ be a Lie group and $\mathfrak{H}$ a subgroup of $\mathfrak{G}$. Let $\pi: \mathfrak{G} \to \mathfrak{G}/\mathfrak{H}$ be the natural mapping. We may put a topology on $\mathfrak{G}/\mathfrak{H}$ by calling a set $U \subset \mathfrak{G}/\mathfrak{H}$ open if $\pi^{-1}(U)$ is open in $\mathfrak{G}$. Under this topology π is a continuous open mapping.

Lemma 9-2. *Let $\mathfrak{H}$ be a connected closed (Lie) subgroup of the Lie group $\mathfrak{G}$. Let $\pi: \mathfrak{G} \to \mathfrak{G}/\mathfrak{H}$ be the natural mapping. Then there is a submanifold M of $\mathfrak{G}$ such that π maps M homeomorphically onto a neighborhood of $\pi(\mathfrak{H})$. Furthermore*

$$\textit{dimensions of } M = \textit{dimension of } \mathfrak{G} - \textit{dimension of } \mathfrak{H}$$

Proof. Let X be the left-invariant vector field on $\mathfrak{G}$ generated by the tangent space to $\mathfrak{H}$ at the identity element. Let V be the coordinate neighborhood whose existence is asserted in Lemma 9-1. We may assume that V has the form

$$V = \{(x_1, \ldots, x_n) \mid x_1^2 + \cdots + x_n^2 < r^2\}$$

Let V_2 be a neighborhood of the identity element in $\mathfrak{G}$ of similar form such that $V_2V_2 \subset V$. Let V_1 be another neighborhood of the identity element such that $V_1^{-1}V_1 \subset V_2$. Let $\theta: V \to V$ be given by

$$\theta(x_1, \ldots, x_n) = (0, \ldots, 0, x_{p+1}, \ldots, x_n)$$

We shall prove that, if $\pi(g_1) = \pi(g_2)$ and $g_1, g_2 \in V_1$, then

$$\theta(g_1) = \theta(g_2)$$

Let $h = g_2^{-1}g_1$. Then $h \in \mathfrak{H} \cap V_2$. Since $\mathfrak{H} \cap V_2$ is connected, $g_2(\mathfrak{H} \cap V_2)$ is a connected subset of $g_2\mathfrak{H} \cap V$. Now $g_2\mathfrak{H}$ is a maximal integral manifold for X, and so, by the argument given in the proof of Lemma 9-1,

$$g_2\mathfrak{H} \cap V = \bigcup_{g \in \theta(g_2\mathfrak{H} \cap V)} \theta^{-1}(g)$$

The sets $\theta^{-1}(g)$ are both open and closed in $g_2\mathfrak{H} \cap V$, and so $g_2(\mathfrak{H} \cap V_2) \subset \theta^{-1}(\theta(g_2))$. Since $g_1 = g_2h \in g_2(\mathfrak{H} \cap V_2)$, we find that $\theta(g_1) = \theta(g_2)$.

Let $M = \theta(V_1)$. Then M is the submanifold of V_1 defined by the equations $x_1 = \cdots = x_p = 0$. The restriction of the mapping π to M is one-to-one and maps M onto $\pi(V_1)$. Since π is an open mapping, $\pi(V_1)$ is a neighborhood of $\pi(\mathfrak{H})$, and the lemma is proved.

We may use Lemma 9-2 to put a differentiable structure on $\mathfrak{G}/\mathfrak{H}$. Let $\pi: \mathfrak{G} \to \mathfrak{G}/\mathfrak{H}$ be the natural mapping and M be the submanifold of $\mathfrak{G}$ whose existence is asserted in Lemma 9-2. If $g \in \mathfrak{G}$, we may use π, which maps gM homeomorphically onto a neighborhood U of $\pi(g)$ in $\mathfrak{G}/\mathfrak{H}$, to transfer the differentiable structure of gM to U. Clearly, the coordinate neighborhoods that are produced in this manner are compatible with one another and define a differentiable structure on $\mathfrak{G}/\mathfrak{H}$. This differentiable structure is characterized by the property that a real-valued function f on $\mathfrak{G}/\mathfrak{H}$ is differentiable at the point $\pi(g)$ if and only if $f \circ \pi$ is differentiable at g. In particular, π is a differentiable mapping, and the reader may readily verify that it is also locally trivial. The results may be summarized in the following theorem.

Theorem 9-4. *If $\mathfrak{H}$ is a connected closed (Lie) subgroup of a Lie group $\mathfrak{G}$, then $\mathfrak{G}$ is a principal bundle with the group $\mathfrak{H}$ under the natural action of $\mathfrak{H}$ on $\mathfrak{G}$ on the right.*

The condition that $\mathfrak{H}$ be connected can be removed from the hypotheses of this theorem. In fact, it is possible to prove the following more general theorem by similar methods. Although we shall refer to this result, the proof will be omitted.

Theorem 9-5. *If $\mathfrak{H}_1$ and $\mathfrak{H}_2$ are closed subgroups of the Lie group $\mathfrak{G}$ such that $\mathfrak{H}_2 \subset \mathfrak{H}_1$, then $\mathfrak{H}_1$ acts naturally on $\mathfrak{G}$ on the right and on $\mathfrak{H}_1/\mathfrak{H}_2$ on the left and gives rise to a fiber bundle whose bundle space may be identified with $\mathfrak{G}/\mathfrak{H}_2$.*

9-6 TRANSITIVE GROUPS

Let $\mathfrak{G}$ be a Lie group. We shall extend the notion of the action of $\mathfrak{G}$ on a manifold. Let S be a topological space. We shall say that $\mathfrak{G}$ is a *continuous left transformation group* on S or that $\mathfrak{G}$ *acts on S on the left* if the following conditions are satisfied:

1. With each $g \in \mathfrak{G}$ there is associated a homeomorphism $s \to gs$ of S.

2. The mapping $(g, s) \to gs$ of $\mathfrak{G} \times S$ into S is continuous.
3. The identity element of $\mathfrak{G}$ is the identity homeomorphism of S.
4. $(g_1g_2)s = g_1(g_2s)$ for $g_1, g_2 \in \mathfrak{G}$ and $s \in S$.

We shall say that $\mathfrak{G}$ acts *transitively* on S if for every pair of points s_1, s_2 in S there is a g in $\mathfrak{G}$ such that $s_2 = gs_1$.

Let the Lie group $\mathfrak{G}$ act transitively on the topological space S on the left. Let $s_0 \in S$. Then the set $\mathfrak{H}$ of all elements of $\mathfrak{G}$ which leave s_0 fixed is a closed subgroup of $\mathfrak{G}$. We may form the homogeneous space $\mathfrak{G}/\mathfrak{H}$. The mapping ψ: $\mathfrak{G}/\mathfrak{H} \to S$ given by $\psi(g\mathfrak{H}) = gs_0$ is well-defined, one-to-one, and onto S. Let π: $\mathfrak{G} \to \mathfrak{G}/\mathfrak{H}$ be the natural mapping, and let μ: $\mathfrak{G} \to S$ be defined by $\mu(g) = gs_0$. Then $\psi \circ \pi = \mu$. Since π is an open mapping, ψ is continuous. The necessary and sufficient condition that ψ be a homeomorphism is that μ: $\mathfrak{G} \to S$ be an open mapping. Because of the transitivity of $\mathfrak{G}$, this condition is satisfied for each mapping $g \to gs$ once it is satisfied for $g \to gs_0$. If ψ is a homeomorphism, we may use it to transfer the differentiable structure of $\mathfrak{G}/\mathfrak{H}$ to S. In this manner $\mathfrak{G}$ induces a differentiable structure on S.

More generally, let $\mathfrak{G}$ be a Lie group and S a set. Assume that for each $g \in \mathfrak{G}$ there is a one-to-one mapping $s \to gs$ of S onto S. And assume that these mappings satisfy the condition $(g_1g_2)s = g_1(g_2s)$. Under these circumstances we shall say that $\mathfrak{G}$ is a Lie group of *permutations* of S. If $\mathfrak{G}$ acts transitively on S, we topologize S in the following manner. Let $s_0 \in S$. The open sets in S are the images of the open sets in $\mathfrak{G}$ under the mapping $g \to gs_0$. This topology does not depend on the choice of the point s_0.

Under the topology which has been induced on S by $\mathfrak{G}$, the reader may verify that $\mathfrak{G}$ becomes a continuous transformation group on S which satisfies the condition that $g \to gs_0$ is an open mapping. We have therefore proved the following theorem.

Theorem 9-6. *If $\mathfrak{G}$ is a Lie group of permutations of a set S which acts transitively on S, then $\mathfrak{G}$ induces a differentiable structure on S.*

Now we shall assume that S is a differentiable manifold and that the Lie group $\mathfrak{G}$ acts transitively on S as a differentiable transformation group. Once again we introduce the projection π: $\mathfrak{G} \to \mathfrak{G}/\mathfrak{H}$, the mapping μ: $\mathfrak{G} \to S$ defined by $\mu(g) = gs_0$, and the mapping ψ: $\mathfrak{G}/\mathfrak{H} \to S$ such that $\mu = \psi \circ \pi$. Since μ is differentiable, it follows that ψ is differentiable. Suppose that ψ is a diffeomorphism. For the induced linear transformations of the tangent spaces we obtain $\mu_* = \psi_* \circ \pi_*$. Both π_* and ψ_* are mappings onto, and so μ_* is onto.

Conversely, we shall prove that, if μ_* is onto, ψ is a diffeomorphism.

In the first place, ψ_* must be onto. Let $s \in S$. By Theorem 2-6, there is a differentiable mapping $\rho\colon U \to \mathfrak{G}/\mathfrak{H}$ with $s \in U$ such that $\psi \circ \rho =$ identity on U. But then $\psi^{-1} = \rho$ on U; hence ψ^{-1} is differentiable at s.

When the mapping $\mu\colon \mathfrak{G} \to S$ satisfies the condition that the induced linear transformation $\mu_*: T(g) \to T(gs)$ is onto, we say that $\mathfrak{G}$ acts with *maximal rank* at s. If $\mathfrak{G}$ acts transitively on S, then $\mathfrak{G}$ acts with maximal rank at each point of S if and only if $\mathfrak{G}$ acts with maximal rank at one point. In this case we merely say that $\mathfrak{G}$ acts with maximal rank on S. In these terms we have proved the following theorem.

Theorem 9-7. *If the Lie group $\mathfrak{G}$ acts transitively on a differentiable manifold M, then the differentiable structure which $\mathfrak{G}$ induces on M is the same as the differentiable structure of the manifold M if and only if $\mathfrak{G}$ acts with maximal rank on M.*

The theory of transitive transformation groups gives rise to interesting examples of differentiable manifolds.

1. *The n-sphere.* Let $O(n + 1)$ be the orthogonal group on V^{n+1}. Let S^n denote the n-sphere. Then $O(n + 1)$ acts transitively and with maximal rank on S^n in a natural way. Let $x \in S^n$. Then the subgroup of elements of $O(n + 1)$ which leave x fixed may be identified with $O(n)$. It follows, from Theorem 9-7, that S^n and $O(n + 1)/O(n)$ have the same differentiable structure.

2. *Stiefel manifolds.* An ordered set of n mutually orthogonal unit vectors in V^{n+p} will be called an n-frame in V^{n+p}. Denote the set of all n-frames in V^{n+p} by $S_{p,n}$. Then the orthogonal group $O(n + p)$ is a transitive Lie group of permutations of $S_{p,n}$. By Theorem 9-6, the Lie group $O(n + p)$ induces a differentiable structure on $S_{p,n}$. The differentiable manifold $S_{p,n}$ is called a *Stiefel manifold.*

The subgroup of $O(n + p)$ that keeps a particular n-frame fixed can be identified with the orthogonal group $O(p)$. Consequently $S_{p,n}$ has the differentiable structure of $O(n + p)/O(p)$. In particular, the m-sphere S^m is the Stiefel manifold $S_{m,1}$.

The set of all unit vectors in V^{n+p} constitutes the $(n + p - 1)$-sphere S^{n+p-1}. We may map $S_{p,n}$ onto S^{n+p-1} by mapping each n-frame $(v_1, \ldots, v_n)$ onto its first vector v_1. The remaining vectors $(v_2, \ldots, v_n)$ constitute an $(n - 1)$-frame in the tangent space to S^{n+p-1} at v_1. Thus $S_{p,n}$ is made up of Stiefel manifolds $S_{p,\,n-1}$ of tangent vectors to S^{n+p-1}, one such manifold at each point of S^{n+p-1}. In other words, $S_{p,n}$ is a fiber bundle over S^{n+p-1} with group $O(n + p - 1)$ and fiber $S_{p,\,n-1}$. In particular, $S_{p,2}$ is the manifold of all unit tangent vectors to S^{p+1}.

3. *Grassman manifolds.* An n-dimensional subspace of V^{n+p} will be called an n-plane. Let $G_{p,n}$ be the set of all n-planes in V^{n+p}. Then the orthogonal group $O(n+p)$ is a transitive Lie group of permutations of $G_{p,n}$. By Theorem 9-6, the Lie group $O(n+p)$ induces a differentiable structure of $G_{p,n}$. With this structure, $G_{p,n}$ is called a *Grassman manifold.*

Let $\mathrm{P}^n \in G_{p,n}$. Let P^p be the orthogonal complement of P^n in V^{n+p}. Then P^p is a p-plane in V^{n+p}. If $\rho \in O(n+p)$, then ρ is determined by its restrictions to P^n and P^p. Suppose that ρ keeps P^n fixed. Then it must also keep P^p fixed. Its restrictions to P^n and P^p may therefore be identified with elements of $O(n)$ and $O(p)$, respectively. Hence the subgroup of $O(n+p)$ that keeps P^n fixed may be identified with $O(n) \times O(p)$. $G_{p,n}$ therefore has the differentiable structure of $O(n+p)/(O(n) \times O(p))$.

Since $O(p)$ may be regarded as a subgroup of $O(n) \times O(p)$, Theorem 9-5 tells us that $S_{p,n}$ is the bundle space of a fiber bundle with group $O(n) \times O(p)$, fiber $O(n)$, and base space $G_{p,n}$. In geometrical terms, the projection of $S_{p,n}$ onto $G_{p,n}$ is the mapping that carries an n-frame onto the n-plane spanned by the vectors of that n-frame.

9-7 REDUCTION OF THE GROUP OF THE BUNDLE

Let $\mathfrak{P}$ be a principal bundle with group $\mathfrak{G}$. Let $\mathfrak{H}$ be a closed subgroup of $\mathfrak{G}$. If $\mathfrak{P}$ is represented by a principal coordinate bundle $\{P, X, \mathfrak{G}, \mathfrak{G}, \{U_\alpha, g_{\alpha\beta}\}\}$ such that $g_{\alpha\beta}(U_\alpha \cap U_\beta) \subset \mathfrak{H}$ for all α, β, we shall say that *the group of* $\mathfrak{P}$ *can be reduced to* $\mathfrak{H}$. We shall obtain a characterization of this property which does not depend upon coordinate transformations.

In general, if $\mathfrak{P}$ is a principal bundle with group $\mathfrak{G}$ and if $\mathfrak{H}$ is a closed subgroup of $\mathfrak{G}$, then $\mathfrak{G}$ acts on the homogeneous space $\mathfrak{G}/\mathfrak{H}$ on the left by the rule $g_1(g_2\mathfrak{H}) = (g_1g_2)\mathfrak{H}$. This action may not be effective, and so it may be necessary to replace $\mathfrak{G}$ by the group $\mathfrak{G}/\mathfrak{H}_0$, where

$$\mathfrak{H}_0 = \bigcap_{g \in \mathfrak{G}} g\mathfrak{H}g^{-1}$$

the largest normal subgroup of $\mathfrak{G}$ which is contained in $\mathfrak{H}$. In this case, the coordinate transformations of $\mathfrak{P}$ have to be replaced by their cosets modulo $\mathfrak{H}_0$ and the bundle manifold P of $\mathfrak{P}$ must be replaced by $P/\mathfrak{H}_0$. However, the base manifold of this new principal bundle is still $P/\mathfrak{G}$. Also the homogeneous spaces $\mathfrak{G}/\mathfrak{H}$ and $(\mathfrak{G}/\mathfrak{H}_0)/(\mathfrak{H}/\mathfrak{H}_0)$ are the same. When this is done, one obtains a fiber bundle $\mathfrak{B}$ with the same base manifold as $\mathfrak{P}$, with group $\mathfrak{G}/\mathfrak{H}_0$, and with fiber $\mathfrak{G}/\mathfrak{H}$. The bundle manifold of $\mathfrak{B}$ may be identified with $P/\mathfrak{H}$ under the natural homeomorphism

$$(z\mathfrak{H}_0, g\mathfrak{H})\mathfrak{G}/\mathfrak{H}_0 \to zg\mathfrak{H}$$

of $[(P/\mathfrak{H}_0) \times (\mathfrak{G}/\mathfrak{H})]/(\mathfrak{G}/\mathfrak{H}_0)$ onto $P/\mathfrak{H}$. $\mathfrak{B}$ will be called the fiber bundle determined by P and $\mathfrak{G}/\mathfrak{H}$.

Lemma 9-3. *If $\mathfrak{P}$ is a principal bundle with bundle manifold P and group $\mathfrak{G}$ and if $\mathfrak{H}$ is a closed subgroup of $\mathfrak{G}$, then $\mathfrak{P}$ is also a principal bundle over $P/\mathfrak{H}$ with group $\mathfrak{H}$.*

Proof. Let $\{U_\alpha, \eta_\alpha\}$, $\eta_\alpha\colon U_\alpha \times \mathfrak{G} \to P$, be a system of coordinate neighborhoods for a representative coordinate bundle for $\mathfrak{P}$. We know that the same covering of the base space will do for a system $\{U_\alpha, \zeta_\alpha\}$, $\zeta_\alpha\colon U_\alpha \times \mathfrak{G}/\mathfrak{H} \to P/\mathfrak{H}$, of coordinate neighborhoods for a representative coordinate bundle for $\mathfrak{B}$, the fiber bundle determined by $\mathfrak{P}$ and $\mathfrak{G}/\mathfrak{H}$. Let $\tau\colon W \to \mathfrak{G}$ be a local cross section of $\mathfrak{G}$, as a fiber bundle over $\mathfrak{G}/\mathfrak{H}$, where W is a neighborhood of the coset $\mathfrak{H}$ in $\mathfrak{G}/\mathfrak{H}$. Our object is to demostrate that the projection $\pi\colon P \to P/\mathfrak{H}$ is locally trivial.

Let $u_0 \in P/\mathfrak{H}$. We may assume that $u_0 = \zeta_\alpha(x_0, g_0\mathfrak{H})$. Let $V = \zeta_\alpha(U_\alpha \times g_0 W)$. Then V is a neighborhood of u_0 in $P/\mathfrak{H}$. We shall define a local cross section $\sigma\colon V \to P$ by

$$\sigma[\zeta_\alpha(x, g\mathfrak{H})] = \eta_\alpha(x, g_0\tau(g_0^{-1}g\mathfrak{H}))$$

Then the mapping $(u, h) \to \sigma(u)h$ is a homeomorphism of $V \times \mathfrak{H}$ into P, which completes the proof.

Theorem 9-8. *If $\mathfrak{P}$ is a principal bundle with group $\mathfrak{G}$ and if $\mathfrak{H}$ is a closed subgroup of $\mathfrak{G}$, then the group of $\mathfrak{P}$ can be reduced to $\mathfrak{H}$ if and only if the fiber bundle $\mathfrak{B}$ determined by $\mathfrak{P}$ and $\mathfrak{G}/\mathfrak{H}$ has a cross section.*

Proof. 1. Assume that the group of $\mathfrak{P}$ can be reduced to $\mathfrak{H}$. Let $\mathfrak{P}$ be represented by a coordinate bundle $\{P, X, \mathfrak{G}, \mathfrak{G}, \{U_\alpha, g_{\alpha\beta}\}\}$ with $g_{\alpha\beta}(U_\alpha \cap U_\beta) \subset \mathfrak{H}$. Then $\mathfrak{B}$ is represented by $\{P/\mathfrak{H}, X, \mathfrak{G}/\mathfrak{H}, \mathfrak{G}/\mathfrak{H}_0, \{U_\alpha, g_{\alpha\beta}\mathfrak{H}_0\}\}$. Let $\eta_\alpha\colon U_\alpha \times \mathfrak{G}/\mathfrak{H} \to P/\mathfrak{H}$ be the coordinate mapping. Define $\sigma_\alpha\colon U_\alpha \to P/\mathfrak{H}$ by $\sigma_\alpha(x) = \eta_\alpha(x, \mathfrak{H})$. Then σ_α is a local cross section of $\mathfrak{B}$. Since

$$\sigma_\beta(x) = \eta_\beta(x, \mathfrak{H}) = \eta_\alpha(x, g_{\beta\alpha}(x)\mathfrak{H}) = \eta_\alpha(x, \mathfrak{H}) = \sigma_\alpha(x)$$

for $x \in U_\alpha \cap U_\beta$, it follows that we may define a cross section σ: $X \to P/\mathfrak{H}$ by $\sigma(x) = \sigma_\alpha(x)$ for $x \in U_\alpha$.

2. Conversely, let $\sigma\colon X \to P/\mathfrak{H}$ be a cross section of $\mathfrak{B}$. Since $\mathfrak{P}$ is a principal bundle over $P/\mathfrak{H}$ with group $\mathfrak{H}$, it is represented by a coordinate bundle with the coordinate covering $\{V_\alpha, \zeta_\alpha\}$, $\zeta_\alpha\colon V_\alpha \times \mathfrak{H} \to P$.

Let $U_\alpha = \sigma^{-1}(V_\alpha)$. Then σ is a diffeomorphism of U_α onto V_α and $\{U_\alpha\}$ is an open covering of X. Define $\tau_\alpha\colon U_\alpha \to P$ by

$$\tau_\alpha[\sigma^{-1}(z)] = \zeta_\alpha(z, e)$$

for $z \in V_\alpha$. Then τ_α is a local cross section of $\mathfrak{P}$ which satisfies the condition that the mapping $(x, g) \to \tau_\alpha(x)g$ is a diffeomorphism of $U_\alpha \times \mathfrak{G}$ into P. We may therefore take $\{U_\alpha, \eta_\alpha\}$, $\eta_\alpha(x, g) = \tau_\alpha(x)g$, as a coordinate covering of X for $\mathfrak{P}$.

The coordinate transformations $h_{\alpha\beta}\colon V_\alpha \cap V_\beta \to \mathfrak{H}$ for $\mathfrak{P}$ as a principal bundle over $P/\mathfrak{H}$ satisfy $\tau_\beta(x) = \tau_\alpha(x)h_{\alpha\beta}\circ\sigma(x)$ when $x \in U_\alpha \cap U_\beta$. Thus $\{h_{\alpha\beta}\circ\sigma\}$ are the coordinate transformations for $\{U_\alpha, \eta_\alpha\}$, and we see that the group of $\mathfrak{P}$ can be reduced to $\mathfrak{H}$.

Let M be a differentiable manifold of dimension n. Assume that M has a Riemann metric. Let $\{U_\alpha\}$ be a system of coordinate neighborhoods of M. We may use the Riemann metric to produce a set $\{X_1^\alpha, \ldots, X_n^\alpha\}$ of n orthonormal differentiable vector fields on U_α. Let $\mathfrak{P}(M)$ be the principal bundle of M. Define $\eta_\alpha\colon U_\alpha \times GL(n) \to \mathfrak{P}(M)$ by

$$\eta_\alpha(x, (a_{ij})) = (x, \Sigma a_{1j}X_j^\alpha, \ldots, \Sigma a_{nj}X_j^\alpha)$$

Then $\{U_\alpha, \eta_\alpha\}$ is a system of coordinate neighborhoods for $\mathfrak{P}(M)$. If $x \in U_\alpha \cap U_\beta$, the linear transformation from the n-frame $(X_1^\alpha(x), \ldots, X_n^\alpha(x))$ to the n-frame $(X_1^\beta(x), \ldots, X_n^\beta(x))$ is orthogonal. Consequently, with respect to this system of coordinate neighborhoods, the coordinate transformations of $\mathfrak{P}(M)$ have values in the orthogonal group $O(n)$. In brief, the group of the principal bundle $\mathfrak{P}(M)$ can be reduced to the orthogonal group.

Conversely, let $\mathfrak{P}(M)$ be the principal bundle of the differentiable manifold M, and assume that the group of $\mathfrak{P}(M)$ can be reduced to the orthogonal group. Let $\{U_\alpha, \eta_\alpha\}$, $\eta_\alpha\colon U_\alpha \times GL(n) \to \mathfrak{P}(M)$, be a system of coordinate neighborhoods for $\mathfrak{P}(M)$ for which the coordinate transformations have values in $O(n)$. Let the n-frame $\{X_1^\alpha(x), \ldots, X_n^\alpha(x)\}$ be given by

$$\eta_\alpha(x, \text{identity}) = (x, X_1^\alpha(x), \ldots, X_n^\alpha(x))$$

for each x in U_α. There is a unique Riemann metric on U_α for which $X_1^\alpha, \ldots, X_n^\alpha$ are orthonormal vector fields. By assumption, if $x \in U_\alpha \cap U_\beta$, the linear transformation from $(X_1^\alpha(x), \ldots, X_n^\alpha(x))$ to $(X_1^\beta(x), \ldots, X_n^\beta(x))$ is orthogonal. Hence the Riemann metrics on U_α and on U_β agree on $U_\alpha \cap U_\beta$, and we obtain a Riemann metric over all of M.

We have therefore proved the following statement.

Theorem 9-9. *The differentiable manifold M has a Riemann metric if and only if the group of its principal bundle $\mathfrak{P}(M)$ can be reduced to the orthogonal group.*

By virtue of Theorem 5-8, the principal bundle of a differentiable manifold can always be reduced to the orthogonal group.

In a similar fashion we may prove that a differentiable manifold M of dimension n has n linearly independent vector fields if and only if the group of its principal bundle $\mathfrak{P}(M)$ can be reduced to the identity. We also leave to the reader the proof that this property is always true of a Lie group.

9-8 VECTOR SPACE BUNDLES

We shall now restrict our attention to fiber bundles whose fiber is V^n, for some n, and whose group is $GL(n)$. Many of the ideas that we shall introduce are capable of more general treatment, which we shall leave for the reader to develop if he is interested. The tangent bundles and cotangent bundles of differentiable manifolds are among these vector space bundles, and so it should not be surprising that this class of fiber bundles plays a fundamental role in the theory of differential topology.

If $\mathfrak{B}$ is a fiber bundle with fiber V^n and group $GL(n)$, we shall denote it by $\mathfrak{B}^n$. The bundle manifold and base manifold of $\mathfrak{B}^n$ will be denoted by $B(\mathfrak{B}^n)$ and $X(\mathfrak{B}^n)$, respectively. If $\pi\colon B(\mathfrak{B}^n) \to X(\mathfrak{B}^n)$ is the projection and if $x \in X(\mathfrak{B}^n)$, the set $\pi^{-1}(\{x\})$ will be called the *fiber over* x. Denote the fiber over x by F_x. We shall see that F_x has the algebraic structure of an n-dimensional vector space. Let $\{U_\alpha, \eta_\alpha\}$, $\eta_\alpha\colon U_\alpha \times V^n \to B(\mathfrak{B}^n)$, be a covering of $B(\mathfrak{B}^n)$ by coordinate neighborhoods. If $x \in U_\alpha$, then we wish to define

$$a_1\eta_\alpha(x, v_1) + a_2\eta_\alpha(x, v_2) = \eta_\alpha(x, a_1v_1 + a_2v_2)$$

for numbers a_1 and a_2. It is easily seen that this definition does not depend on the choice of α, and it gives to F_x the desired structure. The mapping $x \to \eta_\alpha(x, 0)$ for $x \in U_\alpha$ is also well-defined and provides us with a cross section of $\mathfrak{B}^n$ which we shall call the *zero section*.

Let $\mathfrak{B}_1{}^m$ and $\mathfrak{B}_2{}^n$ be two vector space bundles. Let $\pi_1\colon B(\mathfrak{B}_1{}^m) \to X(\mathfrak{B}_1{}^m)$ and $\pi_2\colon B(\mathfrak{B}_2{}^n) \to X(\mathfrak{B}_2{}^n)$ be the corresponding projections. Let $\psi\colon B(\mathfrak{B}_1{}^m) \to B(\mathfrak{B}_2{}^n)$ be a differentiable mapping for which $\pi_1(z_1') = \pi_1(z_1)$ implies $\pi_2\circ\psi(z_1') = \pi_2\circ\psi(z_1)$. That is to say, if $x \in X(\mathfrak{B}_1{}^m)$ and if F_x is the fiber over x, then $\pi_2\circ\psi(F_x)$ is a single point

$\psi_X(x)$ in $X(\mathfrak{B}_2{}^n)$. Thus ψ induces a mapping ψ_X: $X(\mathfrak{B}_1{}^m) \to X(\mathfrak{B}_2{}^n)$ which satisfies $\psi_X \circ \pi_1 = \pi_2 \circ \psi$. If the restriction of ψ to F_x, which we know carries F_x into the fiber over $\psi_X(x)$, is a linear transformation for all $x \in X(\mathfrak{B}_1{}^m)$, then ψ will be called a *linear mapping* of $\mathfrak{B}_1{}^m$ into $\mathfrak{B}_2{}^n$, and we shall write ψ: $\mathfrak{B}_1{}^m \to \mathfrak{B}_2{}^n$.

If $m = n$ and the restriction of ψ to F_x is of rank n for all $x \in X(\mathfrak{B}_1{}^n)$, then ψ: $\mathfrak{B}_1{}^n \to \mathfrak{B}_2{}^n$ will be called a *bundle mapping*. On the other hand, if $X(\mathfrak{B}_1{}^m) = X(\mathfrak{B}_2{}^n)$ and ψ_X: $X(\mathfrak{B}_1{}^m) \to X(\mathfrak{B}_2{}^n)$ is the identity mapping, then ψ: $\mathfrak{B}_1{}^m \to \mathfrak{B}_2{}^n$ will be called a *homomorphism*. A homomorphism is regular if and only if it is one-to-one. Two vector space bundles $\mathfrak{B}_1{}^m$ and $\mathfrak{B}_2{}^n$ with $X(\mathfrak{B}_1{}^m) = X(\mathfrak{B}_2{}^n)$ are equivalent, that is, represent the same fiber bundle, if and only if there is a linear mapping ψ: $\mathfrak{B}_1{}^m \to \mathfrak{B}_2{}^n$ which is both a bundle mapping and a homomorphism.

Lemma 9-4. *Let $\mathfrak{B}^n$ be a vector space bundle with the projection π: $B(\mathfrak{B}^n) \to X(\mathfrak{B}^n)$. For each differentiable manifold M and differentiable mapping θ: $M \to X(\mathfrak{B}^n)$ there is a vector space bundle $\mathfrak{B}_1{}^n$ with*

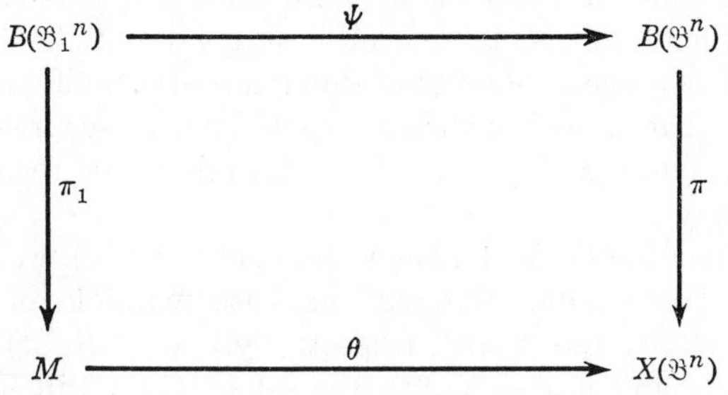

FIG. 9-4

$X(\mathfrak{B}_1{}^n) = M$ and a bundle mapping ψ: $\mathfrak{B}_1{}^n \to \mathfrak{B}^n$ such that $\theta \circ \pi_1 = \pi \circ \psi$, where π_1: $B(\mathfrak{B}_1{}^n) \to M$ is the projection (see Figure 9-4). *Furthermore, $\mathfrak{B}_1{}^n$ is uniquely determined by θ up to equivalence.*

Proof. Form the product manifold $M \times B(\mathfrak{B}^n)$. Let E be the submanifold of $M \times B(\mathfrak{B}^n)$ which consists of the points (m, u) with $\theta(m) = \pi(u)$. Define π_1: $E \to M$ by $\pi_1(m, u) = m$. Let $\{U_\alpha, \eta_\alpha\}$, η_α: $U_\alpha \times V^n \to B(\mathfrak{B}^n)$, be a coordinate covering for $\mathfrak{B}^n$. Define ζ_α: $\theta^{-1}(U_\alpha) \times V^n \to E$ by $\zeta_\alpha(m, v) = (m, \eta_\alpha(\theta(m), v))$. Then the coordinate covering $\{\theta^{-1}(U_\alpha), \zeta_\alpha\}$ makes E the bundle manifold of a vector space bundle $\mathfrak{B}_1{}^n$ with $X(\mathfrak{B}_1{}^n) = M$ and projection π_1.

Define ψ: $E \to B(\mathfrak{B}^n)$ by $\psi(m, u) = u$. Then ψ is differentiable. If $m \in M$, the fiber of $\mathfrak{B}_1{}^n$ over m is the subset $\{m\} \times \pi^{-1}(\{\theta(m)\})$ of E, and ψ maps this set linearly and one-to-one onto the fiber of $\mathfrak{B}^n$

over $\theta(m)$. Hence ψ is a bundle mapping $\psi: \mathfrak{B}_1^n \rightarrow \mathfrak{B}^n$ which satisfies $\theta \circ \pi_1 = \pi \circ \psi$.

Consider the situation illustrated in Figure 9-5, where ψ_1 and ψ_2 are bundle mappings which satisfy $\theta \circ \pi_1 = \pi \circ \psi_1$ and $\theta \circ \pi_2 = \pi \circ \psi_2$. If $z_1 \in B(\mathfrak{B}_1^n)$, the restriction of ψ_2 to the fiber (of $\mathfrak{B}_2^n$) over $\pi_1(z_1)$ is one-to-one and onto the fiber (of $\mathfrak{B}^n$) over $\theta \circ \pi_1(z_1) = \pi \circ \psi_1(z_1)$. Hence there is a unique element $\rho(z_1)$ of $B(\mathfrak{B}_2^n)$ for which $\psi_2(\rho(z_1)) = \psi_1(z_1)$.

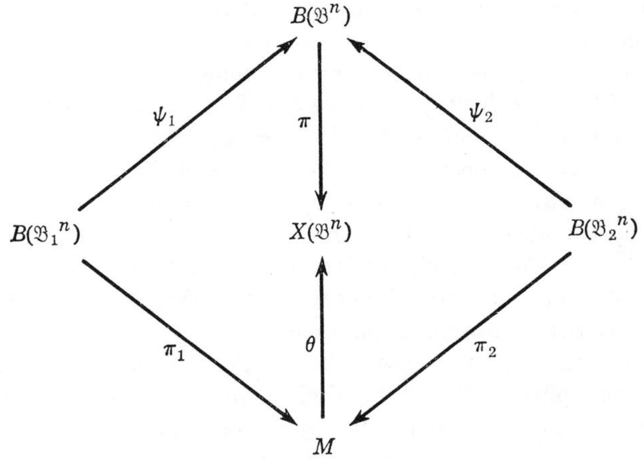

FIG. 9-5

The mapping $\rho: B(\mathfrak{B}_1^n) \rightarrow B(\mathfrak{B}_2^n)$ is one-to-one and onto and satisfies $\psi_2 \circ \rho = \psi_1$. It follows from this relation and the fact that the mappings ψ_1 and ψ_2 induce linear transformations of the fibers of rank n that ρ is differentiable and hence is an equivalence.

The fiber bundle which is represented by $\mathfrak{B}_1^n$ will be called the *bundle induced by* θ.

Theorem 9-10. *If* $\psi: \mathfrak{B}_1^m \rightarrow \mathfrak{B}_2^n$ *is a linear mapping, then there are a homomorphism* $\varphi: \mathfrak{B}_1^m \rightarrow \mathfrak{B}_3^n$ *and a bundle mapping* $\rho: \mathfrak{B}_3^n \rightarrow \mathfrak{B}_2^n$ *such that* $\psi = \rho \circ \varphi$ *and* $\mathfrak{B}_3^n$ *is uniquely determined up to equivalence.*

Proof. Let $\psi_X: X(\mathfrak{B}_1^m) \rightarrow X(\mathfrak{B}_2^n)$ be the mapping induced by ψ on the base manifolds. Let $\mathfrak{B}_3^n$ be the bundle induced by ψ_X and let $\rho: \mathfrak{B}_3^n \rightarrow \mathfrak{B}_2^n$ be the bundle mapping which carries the fiber over x onto the fiber over $\psi_X(x)$. $B(\mathfrak{B}_3^n)$ is a certain submanifold of $X(\mathfrak{B}_1^m) \times B(\mathfrak{B}_2^n)$ and ρ is given by $\rho(x, z_2) = z_2$. Define $\varphi: B(\mathfrak{B}_1^m) \rightarrow X(\mathfrak{B}_1^m) \times B(\mathfrak{B}_2^n)$ by $\varphi(z_1) = (\pi_1(z_1), \psi(z_1))$, where π_1 is the projection of $\mathfrak{B}_1^m$. Then φ is actually a homomorphism of $B(\mathfrak{B}_1^m)$ into $B(\mathfrak{B}_3^n)$ and satisfies $\psi = \rho \circ \varphi$.

Conversely, let $\psi = \rho \circ \varphi$, where $\varphi: \mathfrak{B}_1^m \to \mathfrak{B}^n$ is a homomorphism and ρ is a bundle mapping. Then $\rho_X = \psi_X$, and so $\mathfrak{B}^n$ must be the bundle induced by ψ_X. By Lemma 9-4, $\mathfrak{B}^n$ is uniquely determined up to equivalence.

Let $\mathfrak{B}^n$ be a vector space bundle. Let $Y \subset X(\mathfrak{B}^n)$ be a submanifold of $X(\mathfrak{B}^n)$. Let $\theta: Y \to X(\mathfrak{B}^n)$ be the injection mapping. The bundle induced by θ is called the *restriction* of $\mathfrak{B}^n$ to Y and is denoted by $\mathfrak{B}^n|Y$. The associated bundle mapping $\psi: \mathfrak{B}^n|Y \to \mathfrak{B}^n$ is a diffeomorphism.

Let $\mathfrak{B}_1^m$ and $\mathfrak{B}_2^n$ be two vector space bundles. Form the product manifold $B(\mathfrak{B}_1^m) \times B(\mathfrak{B}_2^n)$. Define $\pi: B(\mathfrak{B}_1^m) \times B(\mathfrak{B}_2^n) \to X(\mathfrak{B}_1^m) \times X(\mathfrak{B}_2^n)$ by $\pi(z_1, z_2) = (\pi_1(z_1), \pi_2(z_2))$, where π_1 and π_2 are the projections of $\mathfrak{B}_1^m$ and $\mathfrak{B}_2^n$, respectively. Then π is a differentiable mapping. The reader may easily demonstrate that $B(\mathfrak{B}_1^m) \times B(\mathfrak{B}_2^n)$ has a natural coordinate bundle structure with the projection π in which the fiber is V^{m+n}. The resulting vector space bundle will be denoted by $\mathfrak{B}_1^m \times \mathfrak{B}_2^n$. The natural projections of $B(\mathfrak{B}_1^m) \times B(\mathfrak{B}_2^n)$ onto $B(\mathfrak{B}_1^m)$ and $B(\mathfrak{B}_2^n)$ are linear mappings.

Consider the special case when $X(\mathfrak{B}_1^m) = X(\mathfrak{B}_2^n) = X$. The diagonal mapping $\delta: X \to X \times X$ given by $\delta(x) = (x, x)$ is a diffeomorphism. The restriction $(\mathfrak{B}_1^m \times \mathfrak{B}_2^n)|\delta(X)$, that is to say, the bundle induced by δ, is called the *Whitney sum* of $\mathfrak{B}_1^m$ and $\mathfrak{B}_2^n$ and is denoted by $\mathfrak{B}_1^m \oplus \mathfrak{B}_2^n$. The fiber of $\mathfrak{B}_1^m \oplus \mathfrak{B}_2^n$ is V^{m+n}. The bundle manifold of $\mathfrak{B}_1^m \oplus \mathfrak{B}_2^n$ may be described as the submanifold of $B(\mathfrak{B}_1^m) \times B(\mathfrak{B}_2^n)$ which consists of those points (z_1, z_2) for which $\pi_1(z_1) = \pi_2(z_2)$. For this model the natural injection of $\mathfrak{B}_1^m \oplus \mathfrak{B}_2^n$ into $\mathfrak{B}_1^m \times \mathfrak{B}_2^n$ is the trivial mapping $(z_1, z_2) \to (z_1, z_2)$. There are obvious natural projections of $\mathfrak{B}_1^m \oplus \mathfrak{B}_2^n$ onto $\mathfrak{B}_1^m$ and $\mathfrak{B}_2^n$. Furthermore, if $\sigma_0: X(\mathfrak{B}_2^n) \to B(\mathfrak{B}_2^n)$ is the zero section of $\mathfrak{B}_2^n$, the mapping $z_1 \to (z_1, \sigma_0 \circ \pi_1(z_1))$ is a one-to-one homomorphism of $\mathfrak{B}_1^m$ into $\mathfrak{B}_1^m \oplus \mathfrak{B}_2^n$ which we shall call the *natural injection.* There is also a natural injection of $\mathfrak{B}_2^n$ into $\mathfrak{B}_1^m \oplus \mathfrak{B}_2^n$. The Whitney sum is a commutative and associative operation on fiber bundles with a common base manifold.

Let $\psi: \mathfrak{B}_1^m \to \mathfrak{B}_2^n$ be a homomorphism. Let $\sigma_0: X(\mathfrak{B}_2^n) \to B(\mathfrak{B}_2^n)$ be the zero section of $\mathfrak{B}_2^n$. The set $\psi^{-1}[\sigma_0(X(\mathfrak{B}_2^n))]$ is called the *kernel* of ψ. If ψ is onto, the rank of the linear transformation that ψ induces on the fiber over x is equal to n for all $x \in X(\mathfrak{B}_1^m)$. In this case the kernel of ψ is a vector space bundle $\mathfrak{K}^{m-n}$ with

$$X(\mathfrak{K}^{m-n}) = X(\mathfrak{B}_1^m)$$

Returning to a general homomorphism $\psi: \mathfrak{B}_1^m \to \mathfrak{B}_2^n$, we may define $z_2 \sim z_2'$ for $z_2, z_2' \in B(\mathfrak{B}_2^n)$ when z_2 and z_2' lie in the same fiber and $z_2' - z_2 \in \psi(B(\mathfrak{B}_1^m))$. The set Q of equivalence classes of elements of $B(\mathfrak{B}_1^m)$ with respect to this relation is called the *cokernel* of ψ. If ψ is one-to-one, Q is the bundle manifold of a vector space bundle $\mathfrak{Q}^{n-m}$ with $X(\mathfrak{Q}^{n-m}) = X(\mathfrak{B}_2^n)$.

Let $\psi: \mathfrak{B}_1^m \to \mathfrak{B}_2^n$ and $\varphi: \mathfrak{B}_2^n \to \mathfrak{B}_3^p$ be homomorphisms such that ψ is one-to-one, φ is onto, and $\psi(B(\mathfrak{B}_1^m)) =$ kernel of φ. In the language of cohomology theory,

$$0 \to \mathfrak{B}_1^m \xrightarrow{\psi} \mathfrak{B}_2^n \xrightarrow{\varphi} \mathfrak{B}_3^p \to 0$$

is an exact sequence. Then there is a homomorphism $\delta: \mathfrak{B}_2^n \to \mathfrak{B}_1^m \oplus \mathfrak{B}_3^p$ which is also a bundle mapping such that $\delta \circ \psi: \mathfrak{B}_1^m \to \mathfrak{B}_1^m \oplus \mathfrak{B}_3^p$ is the natural injection and $\varphi \circ \delta^{-1}: \mathfrak{B}_1^m \oplus \mathfrak{B}_3^p \to \mathfrak{B}_3^p$ is the natural projection.

Let M_1 and M_2 be differentiable manifolds and let $\sigma: M_1 \to M_2$ be a regular differentiable mapping. Let $\mathfrak{T}_1^m$ and $\mathfrak{T}_2^n$ be the tangent bundles of M_1 and M_2, respectively. Let σ_* be the linear transformation that σ induces on the tangent space to M_1. Define $\psi: B(\mathfrak{T}_1^m) \to B(\mathfrak{T}_2^n)$ by $\psi(x, \alpha) = (\sigma(x), \sigma_*(\alpha))$. Then ψ is a linear mapping. By Theorem 9-10, there are a homomorphism $\varphi: \mathfrak{T}_1^m \to \mathfrak{B}^n$ and a bundle mapping $\rho: \mathfrak{B}^n \to \mathfrak{T}_2^n$ such that $\psi = \rho \circ \varphi$. $\mathfrak{B}^n$ is the bundle induced by σ. Since σ is regular, φ is one-to-one. The cokernel of φ is called the *normal bundle* of σ and is denoted by $\mathfrak{N}_\sigma$. It follows that

$$0 \to \mathfrak{T}_1^m \xrightarrow{\varphi} \mathfrak{B}^n \xrightarrow{\tau} \mathfrak{N}_\sigma \to 0$$

is an exact sequence, where τ is the natural mapping. Consequently the bundle induced by σ from $\mathfrak{T}_2^n$ is equivalent to $\mathfrak{T}_1^m \oplus \mathfrak{N}_\sigma$.

9-9 UNIVERSAL BUNDLES

Let $G_{p,n}$ be the Grassmann manifold of n-planes in V^{n+p}. Let B be the submanifold of $G_{p,n} \times V^{n+p}$ consisting of the pairs (H, v), where v is a vector in the n-plane H. We shall describe a bundle structure on B. The projection $\pi: B \to G_{p,n}$ is given by $\pi(H, v) = H$. In Section 9-6 we saw that the Stiefel manifold $S_{p,n}$ of n-frames in V^{n+p} is a fiber bundle over $G_{p,n}$. Hence, if $H_0 \in G_{p,n}$ there is a local cross section $\sigma: U \to S_{p,n}$ with $H_0 \in U$. Let the vectors of the n-frame $\sigma(H)$ be $(u_1(H), \ldots, u_n(H))$. Then H is spanned by $u_1(H), \ldots, u_n(H)$. We may define $\eta: U \times V^n \to B$ by

$$\eta(H, y) = (H, y_1u_1(H) + \cdots + y_nu_n(H))$$

where $y = (y_1, \ldots, y_n)$. Then η is a diffeomorphism. We shall leave to the reader the verification that all such pairs U, η are compatible with one another and make B the bundle manifold of a vector space bundle $\mathfrak{U}_p{}^n$. $\mathfrak{U}_p{}^n$ is called a *universal bundle.* The following theorem illustrates the appropriateness of this name.

Theorem 9-11. *Let $\mathfrak{B}^n$ be a vector space bundle. Then the following two properties of $\mathfrak{B}^n$ are equivalent:*

(i) *There is a vector space bundle $\mathfrak{C}^p$ such that $\mathfrak{B}^n \oplus \mathfrak{C}^p$ is equivalent to $X(\mathfrak{B}^n) \times V^{n+p}$.*

(ii) *There is a universal bundle $\mathfrak{U}_p{}^n$ and a bundle mapping $\psi: \mathfrak{B}^n \to \mathfrak{U}_p{}^n$.*

Proof. Assume property i. Then the natural injection of $\mathfrak{B}^n$ into $\mathfrak{B}^n \oplus \mathfrak{C}^p$ followed by the equivalence of $\mathfrak{B}^n \oplus \mathfrak{C}^p$ and $X(\mathfrak{B}^n) \times V^{n+p}$ provides us with a regular homomorphism $\varphi: \mathfrak{B}^n \to X(\mathfrak{B}^n) \times V^{n+p}$. Thus if F_x is the fiber of $\mathfrak{B}^n$ over x, then $\varphi(F_x)$ is an n-plane in the fiber of $X(\mathfrak{B}^n) \times V^{n+p}$ over x, that is, in V^{n+p} itself. Hence if $\pi: B(\mathfrak{B}^n) \to X(\mathfrak{B}^n)$ is the projection, we may define a mapping $\Phi: B(\mathfrak{B}^n) \to G_{p,n}$ by $\Phi(z) = \varphi(F_{\pi(z)})$. We shall prove that Φ is differentiable.

Let U, η: $U \times V^n \to B(\mathfrak{B}^n)$ be a coordinate neighborhood of $\mathfrak{B}^n$. It suffices to show that the restriction of Φ to $\pi^{-1}(U)$ is differentiable. Let $\{v^1, \ldots, v^n\}$ be a fixed basis of V^n. Define $u^i: \pi^{-1}(U) \to V^{n+p}$ by

$$u^i(z) = \varphi[\eta(\pi(z), v^i)]$$

Then $u^1, \ldots, u^n$ are differentiable mappings and the vectors $u^1(z), \ldots, u^n(z)$ span $\Phi(z)$. It follows that Φ is differentiable.

If $z \in B(\mathfrak{B}^n)$, then $\varphi(z)$ is a vector in the n-plane $\Phi(z)$. Hence $z \to (\Phi(z), \varphi(z))$ is a differentiable mapping of $B(\mathfrak{B}^n)$ into $B(\mathfrak{U}_p{}^n)$. Since φ induces a linear transformation of rank n on the fibers of $\mathfrak{B}^n$, the same is true of $z \to (\Phi(z), \varphi(z))$, which is therefore a bundle mapping.

Conversely, assume property ii. We may identify $B(\mathfrak{U}_p{}^n)$ with a submanifold of $G_{p,n} \times V^{n+p}$. Let θ: $G_{p,n} \times V^{n+p} \to V^{n+p}$ be the natural projection. Let $\varphi = \theta \circ \psi$. Then φ: $B(\mathfrak{B}^n) \to V^{n+p}$ is a differentiable mapping whose restriction to each fiber of $\mathfrak{B}^n$ is a linear transformation of rank n. Let $\pi: B(\mathfrak{B}^n) \to X(\mathfrak{B}^n)$ be the projection. Then define $\tau: B(\mathfrak{B}^n) \to X(\mathfrak{B}^n) \times V^{n+p}$ by

$$\tau(z) = (\pi(z), \varphi(z))$$

τ is a regular homomorphism. We may therefore form the cokernel $\mathfrak{C}^p$ of τ. Since

$$0 \to \mathfrak{B}^n \xrightarrow{\tau} X(\mathfrak{B}^n) \times V^{n+p} \xrightarrow{\nu} \mathfrak{C}^p \to 0$$

is an exact sequence, where ν is the natural mapping, it follows that $X(\mathfrak{B}^n) \times V^{n+p}$ is equivalent to $\mathfrak{B}^n \oplus \mathfrak{C}^p$.

Corollary. *The tangent bundle* $\mathfrak{T}^n(M)$ *of a differentiable manifold* M *of dimension* n *has a bundle mapping* $\psi\colon \mathfrak{T}^n(M) \to \mathfrak{U}^n_{n+1}$.

Proof. By the Whitney imbedding theorem (Theorem 6-3), there is a diffeomorphism $\sigma\colon M \to V^{2n+1}$. Let $\mathfrak{C}^{n+1}$ be the normal bundle of σ, and let $\mathfrak{B}^{2n+1}$ be the bundle induced by σ from the tangent bundle of V^{2n+1}. Then $\mathfrak{B}^{2n+1}$ is equivalent to $\mathfrak{T}^n(M) \oplus \mathfrak{C}^{n+1}$. But $\mathfrak{B}^{2n+1}$ is equivalent to $M \times V^{2n+1}$, and so we may apply Theorem 9-11.

CHAPTER 10

Multilinear Algebra

Let $V_1, \ldots, V_r, W$ be vector spaces. The set of all ordered r-tuples $(\alpha_1, \ldots, \alpha_r)$, where α_i is a vector from V_i, will be denoted by $V_1 \times \cdots \times V_r$. A mapping

$$\varphi: V_1 \times \cdots \times V_r \to W$$

will be called *multilinear* if it satisfies the condition

$$\begin{aligned}\varphi(\alpha_1, \ldots, a\alpha_i + a'\alpha_i', \ldots, \alpha_r) \\ = a\varphi(\alpha_1, \ldots, \alpha_i, \ldots, \alpha_r) + a'\varphi(\alpha_1, \ldots, \alpha_i', \ldots, \alpha_r)\end{aligned}$$

where a and a' are numbers, for $i = 1, \ldots, r$. Roughly speaking, φ is multilinear if it is linear in each of its "variables" separately. Multilinear algebra is the theory of such mappings and certain vector spaces to which they give rise.

We shall denote the set of all multilinear mappings of $V_1 \times \cdots \times V_r$ into W by $L(V_1, \ldots, V_r; W)$. This set becomes a vector space in a natural way if we define addition by

$$(\varphi_1 + \varphi_2)(\alpha_1, \ldots, \alpha_r) = \varphi_1(\alpha_1, \ldots, \alpha_r) + \varphi_2(\alpha_1, \ldots, \alpha_r)$$

and multiplication by numbers by

$$(a\varphi)(\alpha_1, \ldots, \alpha_r) = a\varphi(\alpha_1, \ldots, \alpha_r)$$

We shall use this notation even when $r = 1$. In this case $L(V; W)$ is the vector space of all linear transformations of V into W.

When $r = 2$, the elements of $L(V_1, V_2; W)$ are called bilinear mappings of $V_1 \times V_2$ into W. We shall usually carry out the details of the multilinear algebra only for bilinear mappings, but the reader should bear in mind (and check for himself) that the proofs are also valid in the more general case.

10-1 THE TENSOR PRODUCT OF VECTOR SPACES

Let U and V be vector spaces. A vector space T, together with a bilinear mapping θ: $U \times V \to T$, will be called a *tensor product* of U and V if it satisfies the following conditions.

1. The image set $\theta(U \times V)$ spans T.

2. If φ: $U \times V \to W$ is any bilinear mapping, there is a linear transformation λ: $T \to W$ such that $\varphi = \lambda \circ \theta$ (see Figure 10-1).

Since condition 2 requires that the equation $\varphi = \lambda \circ \theta$ have a solution λ for every bilinear mapping φ, no matter what the vector space W

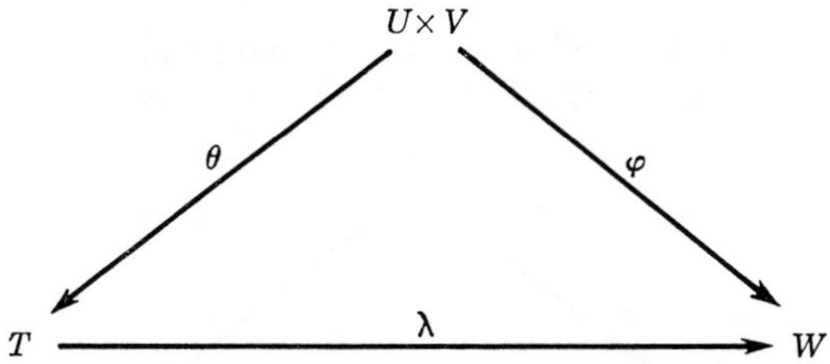

FIG. 10-1

may be, the reader may feel that only very special vector spaces U and V may have a tensor product. Nevertheless, we shall eventually show how to construct a tensor product of any pair of vector spaces U, V. Putting aside, for the present, the question of the existence of tensor products, we shall show how the theory of the bilinear mappings of $U \times V$ is equivalent to the theory of the linear transformations of a tensor product of U and V.

Theorem 10-1. *Let T, θ be a tensor product of the vector spaces U and V. If φ: $U \times V \to W$ is a bilinear mapping, then the linear transformation λ: $T \to W$ such that $\varphi = \lambda \circ \theta$ is uniquely determined by φ. The mapping $\varphi \to \lambda$ is an isomorphism of $L(U, V; W)$ onto $L(T; W)$.*

Proof. If λ_1: $T \to W$ and λ_2: $T \to W$ are linear transformations such that $\lambda_1 \circ \theta = \lambda_2 \circ \theta$, then $\lambda_1 - \lambda_2$ maps the elements of $\theta(U \times V)$ onto zero. It follows from condition 1 for the tensor product that $\lambda_1 - \lambda_2 = 0$. Thus $\varphi \to \lambda$ is a well-defined mapping of $L(U, V; W)$ into $L(T; W)$. On the other hand, if λ is an element of $L(T; W)$, then $\varphi = \lambda \circ \theta$ is an element of $L(U, V; W)$ which maps onto λ, and so the mapping $\varphi \to \lambda$ is onto $L(T; W)$.

The mapping $\varphi \to \lambda$ is clearly one-to-one. It remains to show that it is linear. Let φ_1 and φ_2 be elements of $L(U, V; W)$, and let λ_1 and λ_2 be the elements of $L(T; W)$ such that $\varphi_1 = \lambda_1 \circ \theta$ and $\varphi_2 = \lambda_2 \circ \theta$. If

a_1 and a_2 are numbers, then

$$a_1\varphi_1 + a_2\varphi_2 = a_1\lambda_1\circ\theta + a_2\lambda_2\circ\theta = (a_1\lambda_1 + a_2\lambda_2)\circ\theta$$

It follows that $a_1\varphi_1 + a_2\varphi_2 \to a_1\lambda_1 + a_2\lambda_2$, and so this mapping is linear.

The following theorem shows that there is essentially only one tensor product of a pair of vector spaces.

Theorem 10-2. *Let T, θ and T', θ' be two tensor products of the vector spaces U and V. Then there is an isomorphism σ of T onto T' such that $\theta' = \sigma\circ\theta$.*

Proof. $\theta'\colon U \times V \to T'$ is a bilinear mapping, and so there is a linear transformation $\sigma\colon T \to T'$ such that $\theta' = \sigma\circ\theta$ (compare Figure

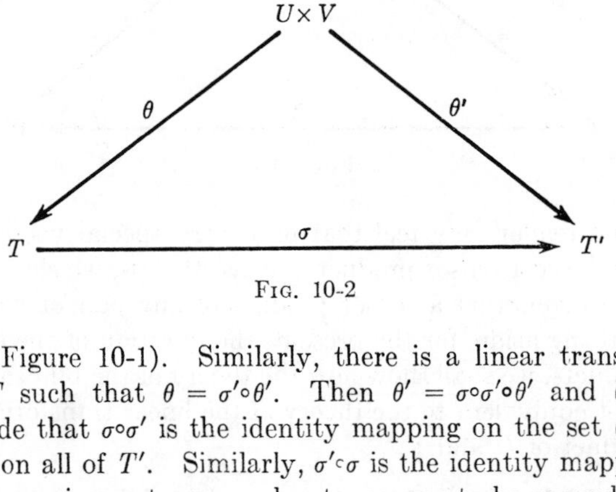

FIG. 10-2

10-2 with Figure 10-1). Similarly, there is a linear transformation $\sigma'\colon T' \to T$ such that $\theta = \sigma'\circ\theta'$. Then $\theta' = \sigma\circ\sigma'\circ\theta'$ and $\theta = \sigma'\circ\sigma\circ\theta$. We conclude that $\sigma\circ\sigma'$ is the identity mapping on the set $\theta'(U \times V)$ and hence on all of T'. Similarly, $\sigma'\circ\sigma$ is the identity mapping on T, and therefore σ is one-to-one and onto, as was to be proved.

We shall now construct a tensor product of the two vector spaces U and V. Let R denote the real numbers, and let Q denote the set of all mappings $f\colon U \times V \to R$ such that $f(\alpha, \beta) = 0$ except for a finite number of elements (α, β) in $U \times V$. Q is readily made into a vector space by defining

$$(a_1f_1 + a_2f_2)(\alpha, \beta) = a_1f_1(\alpha, \beta) + a_2f_2(\alpha, \beta)$$

where a_1 and a_2 are numbers and f_1 and f_2 are elements of Q. It is more convenient and suggestive to write an element f of Q as a formal sum

$$f = \Sigma a_{\alpha,\beta}(\alpha, \beta)$$

over the elements of $U \times V$, where $a_{\alpha,\beta} = f(\alpha, \beta)$. Then the vector space operations take the form

$$c\Sigma a_{\alpha,\beta}(\alpha, \beta) = \Sigma(ca_{\alpha,\beta})(\alpha, \beta)$$
$$\Sigma a_{\alpha,\beta}(\alpha, \beta) + \Sigma b_{\alpha,\beta}(\alpha, \beta) = \Sigma(a_{\alpha,\beta} + b_{\alpha,\beta})(\alpha, \beta)$$

We shall agree to omit, when convenient, those terms of the formal sum $\Sigma a_{\alpha,\beta}(\alpha, \beta)$ whose coefficients $a_{\alpha,\beta}$ are zero and to write (α, β) for the term $1(\alpha, \beta)$ and $-(\alpha, \beta)$ for the term $(-1)(\alpha, \beta)$ when these terms occur.

Let N be the subspace of Q which is spanned by all elements of the form

$$(a_1\alpha_1 + a_2\alpha_2, b_1\beta_1 + b_2\beta_2) - a_1b_1(\alpha_1, \beta_1) - a_1b_2(\alpha_1, \beta_2) - a_2b_1(\alpha_2, \beta_1) - a_2b_2(\alpha_2, \beta_2) \quad (10\text{-}1\text{-}1)$$

Let T be the quotient space Q/N, and let $\theta: Q \to T$ be the natural homomorphism If we identify the element (α, β) of $U \times V$ with the formal sum (α, β) in Q, then $U \times V$ may be regarded as a subset of Q. Under these circumstances, θ may be restricted to a mapping of $U \times V$ into T. We shall see that T, θ is then a tensor product of U and V.

Since the formal sums

$$(a_1\alpha_1 + a_2\alpha_2, \beta) - a_1(\alpha_1, \beta) - a_2(\alpha_2, \beta)$$
$$(\alpha, b_1\beta_1 + b_2\beta_2) - b_1(\alpha, \beta_1) - b_2(\alpha, \beta_2)$$

are in N, it follows that they are carried onto zero by θ. Hence

$$\theta(a_1\alpha_1 + a_2\alpha_2, \beta) = a_1\theta(\alpha_1, \beta) + a_2\theta(\alpha_2, \beta)$$
$$\theta(\alpha, b_1\beta_1 + b_2\beta_2) = b_1\theta(\alpha, \beta_1) + b_2\theta(\alpha, \beta_2)$$

which proves that θ is bilinear on the set $U \times V$. Since the subset $U \times V$ spans Q, it follows that the image set $\theta(U \times V)$ spans T. Let $\varphi: U \times V \to W$ be a bilinear mapping. Then define $\Phi: Q \to W$ by

$$\Phi(\Sigma a_{\alpha,\beta}(\alpha, \beta)) = \Sigma a_{\alpha,\beta}\varphi(\alpha, \beta)$$

The mapping Φ is well-defined since $U \times V$ is a basis of Q. It is clear that Φ is a linear transformation. Since φ is bilinear, Φ maps the elements (10-1-1), which span N, onto zero. Hence Φ maps all of N onto zero, and we may define a linear transformation $\lambda: T \to W$ by

$$\lambda(f + N) = \Phi(f)$$

for all elements f of Q. In particular, $(\lambda \circ \theta)(\alpha, \beta) = \Phi(\alpha, \beta) = \varphi(\alpha, \beta)$ for all elements (α, β) of $U \times V$, and so all the conditions for a tensor product of U and V are satisfied.

The tensor product which we have just constructed will be denoted by $U \otimes V$ and will be called *the* tensor product of U and V. We have seen that all tensor products of U and V are isomorphic to this one.

The element $(\alpha, \beta) + N$ of $U \otimes V$ will be denoted by $\alpha \otimes \beta$. Since the elements of the form $\alpha \otimes \beta$ span $U \otimes V$, we may write every element ξ of $U \otimes V$ in the form

$$\xi = \Sigma a_{\alpha,\beta}\, \alpha \otimes \beta$$

These generators satisfy the rules

$$\begin{aligned}(a_1\alpha_1 + a_2\alpha_2) \otimes \beta &= a_1\alpha_1 \otimes \beta + a_2\alpha_2 \otimes \beta\\ \alpha \otimes (b_1\beta_1 + b_2\beta_2) &= b_1\alpha \otimes \beta_1 + b_2\alpha \otimes \beta_2\end{aligned}$$

It is frequently convenient to define a linear transformation $\sigma: U \otimes V \to W$ by specifying the images $\sigma(\alpha \otimes \beta)$ of the generators $\alpha \otimes \beta$. In spite of the fact that these generators are not linearly independent, we have a simple criterion which enables us to say when such a mapping is well-defined: *Let $\varphi: U \times V \to W$ be a mapping. Then there is a linear transformation $\lambda: U \otimes V \to W$ such that $\lambda(\alpha \otimes \beta) = \varphi(\alpha, \beta)$ for all (α, β) in $U \times V$ if and only if φ is bilinear.* The sufficiency of this criterion results from the fact that $U \otimes V$ is a tensor product of U and V. The necessity comes from the fact that any linear transformation $\lambda: U \otimes V \to W$ induces a linear transformation $\Lambda: Q \to W$ by the rule

$$\Lambda(\xi) = \lambda(\xi + N)$$

This mapping is bilinear when it is restricted to $U \times V$. This criterion makes it unnecessary to check that

$$\Sigma a_{\alpha,\beta}\varphi(\alpha, \beta) = \Sigma b_{\alpha,\beta}\varphi(\alpha, \beta)$$

whenever

$$\Sigma a_{\alpha,\beta}\, \alpha \otimes \beta = \Sigma b_{\alpha,\beta}\, \alpha \otimes \beta$$

EXERCISES

1. Let U, V, W be vector spaces. Prove that $L(U, V; W)$ is isomorphic to $L(U; L(V; W))$.

2. Define a tensor product of r vector spaces $V_1, \ldots, V_r$ and show that it is essentially unique. Prove that if T is a tensor product of $V_1, \ldots, V_r$, then there is an isomorphism of $L(V_1, \ldots, V_r; W)$ onto $L(T; W)$.

3. Construct the tensor product $V_1 \otimes \cdots \otimes V_r$ of the vector spaces $V_1, \ldots, V_r$.

10-2 PROPERTIES OF $U \otimes V$

Theorem 10-3. *Let U and V be vector spaces. There is an isomorphism σ of $U \otimes V$ onto $V \otimes U$ such that $\sigma(\alpha \otimes \beta) = \beta \otimes \alpha$.*

Proof. Since the mapping $(\alpha, \beta) \rightarrow \beta \otimes \alpha$ is bilinear, there is a linear transformation $\sigma: U \otimes V \rightarrow V \otimes U$ such that $\sigma(\alpha \otimes \beta) = \beta \otimes \alpha$. Similarly, there is a linear transformation $\sigma': V \otimes U \rightarrow U \otimes V$ such that $\sigma'(\beta \otimes \alpha) = \alpha \otimes \beta$. Then $\sigma \circ \sigma'$ and $\sigma' \circ \sigma$ are identity mappings, and so σ is one-to-one and onto.

Theorem 10-4. *Let U, V, W be vector spaces. Then there is an isomorphism σ of $(U \otimes V) \otimes W$ onto $U \otimes V \otimes W$ such that*

$$\sigma[(\alpha \otimes \beta) \otimes \gamma] = \alpha \otimes \beta \otimes \gamma$$

Proof. Let γ be an element of W. Since the mapping $(\alpha, \beta) \rightarrow \alpha \otimes \beta \otimes \gamma$ is bilinear, there is a linear transformation $\sigma_\gamma: U \otimes V \rightarrow U \otimes V \otimes W$ such that $\sigma_\gamma(\alpha \otimes \beta) = \alpha \otimes \beta \otimes \gamma$. Since the mapping $(\alpha \otimes \beta, \gamma) \rightarrow \sigma_\gamma(\alpha \otimes \beta) = \alpha \otimes \beta \otimes \gamma$ is bilinear, there is a linear transformation $\sigma: (U \otimes V) \otimes W \rightarrow U \otimes V \otimes W$ such that

$$\sigma[(\alpha \otimes \beta) \otimes \gamma] = \alpha \otimes \beta \otimes \gamma$$

On the other hand, the mapping $(\alpha, \beta, \gamma) \rightarrow (\alpha \otimes \beta) \otimes \gamma$ is multilinear. Hence there is a linear transformation $\sigma': U \otimes V \otimes W \rightarrow (U \otimes V) \otimes W$ such that $\sigma'(\alpha \otimes \beta \otimes \gamma) = (\alpha \otimes \beta) \otimes \gamma$. Since $\sigma \circ \sigma'$ and $\sigma' \circ \sigma$ are identity mappings, σ is an isomorphism.

Of course, there is also an isomorphism σ'' of $U \otimes (V \otimes W)$ onto $U \otimes V \otimes W$ such that $\sigma''[\alpha \otimes (\beta \otimes \gamma)] = \alpha \otimes \beta \otimes \gamma$.

The isomorphisms described by these two theorems are called the *natural* isomorphisms in the situations to which they apply. They enable us to treat tensor multiplication as if it were a commutative and associative operation.

Theorem 10-5. *If R denotes the real numbers, then there is an isomorphism σ of $R \otimes V$ onto V such that $\sigma(a \otimes \alpha) = a\alpha$.*

Proof. The mapping $(a, \alpha) \rightarrow a\alpha$ is bilinear, and so there is a linear transformation $\sigma: R \otimes V \rightarrow V$ such that $\sigma(a \otimes \alpha) = a\alpha$. On the other hand, the mapping $\sigma': V \rightarrow R \otimes V$ defined by $\sigma'(\alpha) = 1 \otimes \alpha$ is linear. Since $(\sigma' \circ \sigma)(a \otimes \alpha) = 1 \otimes a\alpha = a(1 \otimes \alpha) = a \otimes \alpha$, it follows that $\sigma' \circ \sigma$ is the identity mapping on all of $R \otimes V$. Also $\sigma \circ \sigma'$ is the identity mapping on V, and so we conclude that σ is an isomorphism.

Theorem 10-6. *Let U and V be vector spaces of finite dimension. Let $\{\alpha_1, \ldots, \alpha_m\}$ be a basis of U and $\{\beta_1, \ldots, \beta_n\}$ be a basis of V. Then the mn elements $\alpha_1 \otimes \beta_1, \alpha_1 \otimes \beta_2, \ldots, \alpha_1 \otimes \beta_n, \alpha_2 \otimes \beta_1, \ldots, \alpha_m \otimes \beta_n$ constitute a basis of $U \otimes V$.*

Proof. The mapping

$$\Big(\sum_{i=1}^{m} a_i\alpha_i, \sum_{j=1}^{n} b_j\beta_j\Big) \to a_k b_l$$

for fixed integers k and l, is bilinear. Hence there is a linear functional λ_{kl} of $U \otimes V$ such that

$$\lambda_{kl}\Big[\Big(\sum_{i=1}^{m} a_i\alpha_i\Big) \otimes \Big(\sum_{j=1}^{n} b_j\beta_j\Big)\Big] = a_k b_l$$

In particular,

$$\lambda_{kl}(\alpha_i \otimes \beta_j) = \begin{cases} 0 & \text{if } i \neq k \text{ or } j \neq l \\ 1 & \text{if } i = k \text{ and } j = l \end{cases}$$

Suppose that

$$\sum_{i=1}^{m}\sum_{j=1}^{n} a_{ij}\,\alpha_i \otimes \beta_j = 0$$

If we apply λ_{kl} to this relation, we find that $a_{kl} = 0$. Since this is true for $1 \leqq k \leqq m$, $1 \leqq l \leqq n$, we see that the elements $\alpha_1 \otimes \beta_1$, $\alpha_1 \otimes \beta_2, \ldots, \alpha_m \otimes \beta_n$ are linearly independent. On the other hand, every element of $U \otimes V$ of the form $\alpha \otimes \beta$ is a linear combination of these elements; hence they span all of $U \otimes V$. Consequently, they constitute a basis for $U \otimes V$.

Theorem 10-7. *Let V be a finite-dimensional vector space. Let V^* be the dual space to V. Then there is an isomorphism σ of $V \otimes V^*$ onto $L(V; V)$ such that $\tau = \sigma(\alpha \otimes \lambda)$ is the linear transformation*

$$\tau(\beta) = \lambda(\beta)\alpha$$

for all β in V.

Proof. If α is in V and λ is in V^*, define $\tau_{\alpha,\lambda}\colon V \to V$ by

$$\tau_{\alpha,\lambda}(\beta) = \lambda(\beta)\alpha$$

Then the mapping $(\alpha, \lambda) \to \tau_{\alpha,\lambda}$ is bilinear. Hence there is a linear transformation $\sigma\colon V \otimes V^* \to L(V\colon V)$ such that $\sigma(\alpha \otimes \lambda) = \tau_{\alpha,\lambda}$. Let $\{\alpha_1, \ldots, \alpha_n\}$ be a basis of V, and let $\{\lambda_1, \ldots, \lambda_n\}$ be the dual basis of V^*. Since $V \otimes V^*$ and $L(V; V)$ have the same dimension, it suffices to prove that the elements $\tau_{\alpha_1,\lambda_1}, \tau_{\alpha_1,\lambda_2}, \ldots, \tau_{\alpha_n,\lambda_n}$ are linearly independent.

Suppose that

$$\sum_{i=1}^{n}\sum_{j=1}^{n} b_{ij}\tau_{\alpha_i,\lambda_j} = 0$$

Then

$$\sum_{i=1}^{n} b_{ik}\alpha_i = \sum_{i=1}^{n}\sum_{j=1}^{n} b_{ij}\tau_{\alpha_i,\lambda_j}(\alpha_k) = 0$$

Hence $b_{1k} = b_{2k} = \cdots = b_{nk} = 0$. Since this is true for $k = 1, \ldots, n$, the elements $\tau_{\alpha_i, \lambda_j}$ are linearly independent.

EXERCISES

1. Let U, V, W be vector spaces, and let S be the direct sum of U and V. Prove that there is an isomorphism σ of the direct sum of $U \otimes W$ and $V \otimes W$ onto $S \otimes W$ such that

$$\sigma(\alpha \otimes \gamma, \beta \otimes \gamma) = (\alpha, \beta) \otimes \gamma$$

for α in U, β in V, and γ in W. If we denote the direct sum by $\oplus$, then this is the distributive law for the tensor product which says that $(U \oplus V) \otimes W$ is isomorphic to $(U \otimes W) \oplus (V \otimes W)$ in a natural way.

2. Let U be a vector space, and let V be a finite-dimensional vector space. If $\{\beta_1, \ldots, \beta_n\}$ is a basis of V, prove that every element of $U \otimes V$ can be written in one and only one way in the form

$$\alpha_1 \otimes \beta_1 + \cdots + \alpha_n \otimes \beta_n$$

where $\alpha_1, \ldots, \alpha_n$ are from U.

10-3 THE TENSOR PRODUCT OF LINEAR TRANSFORMATIONS

Theorem 10-8. *Let $\sigma\colon U \to V$ and $\sigma'\colon U' \to V'$ be linear transformations. Then there is a linear transformation $\lambda\colon U \otimes U' \to V \otimes V'$ such that $\lambda(\alpha \otimes \alpha') = \sigma(\alpha) \otimes \sigma'(\alpha')$.*

Proof. The mapping $\varphi\colon U \times U' \to V \otimes V'$ defined by

$$\varphi(\alpha, \alpha') = \sigma(\alpha) \otimes \sigma'(\alpha')$$

is bilinear. Hence there is a linear transformation $\lambda\colon U \otimes U' \to V \otimes V'$ such that $\lambda(\alpha \otimes \alpha') = \sigma(\alpha) \otimes \sigma'(\alpha')$.

The mapping $\lambda\colon U \otimes U' \to V \otimes V'$ is called the tensor-product transformation of σ and σ' and will be denoted by $\sigma \otimes \sigma'$. Let $\sigma\colon U \to V$, $\sigma'\colon U' \to V'$, $\tau\colon V \to W$, and $\tau'\colon V' \to W'$. Then

$$\begin{aligned}[(\tau \otimes \tau')\circ(\sigma \otimes \sigma')](\alpha \otimes \alpha') &= (\tau \otimes \tau')[\sigma(\alpha) \otimes \sigma'(\alpha')] \\ &= (\tau\circ\sigma)(\alpha) \otimes (\tau'\circ\sigma')(\alpha') = (\tau\circ\sigma \otimes \tau'\circ\sigma')(\alpha \otimes \alpha')\end{aligned}$$

Consequently

$$(\tau \otimes \tau')\circ(\sigma \otimes \sigma') = \tau\circ\sigma \otimes \tau'\circ\sigma'$$

on all of $U \otimes U'$.

Let U and V be vector spaces, and let U^* and V^* be their dual spaces, respectively. Then λ in U^* and μ in V^* are linear transformations, and we may form their tensor-product transformation $\lambda \otimes \mu\colon U \otimes V \to$ real numbers. Thus $\lambda \otimes \mu$ is in the dual space $(U \otimes V)^*$ of $U \otimes V$. On the other hand, we may also form the tensor product $U^* \otimes V^*$. In order to distinguish $\lambda \otimes \mu$ as an element of $U^* \otimes V^*$ from the product transformation of λ and μ, we shall temporarily denote the tensor-product transformation of the transformations λ and μ by $\lambda \otimes_m \mu$.

Theorem 10-9. *Let U and V be finite-dimensional vector spaces. There is an isomorphism σ of $U^* \otimes V^*$ onto $(U \otimes V)^*$ such that $\sigma(\lambda \otimes \mu)$ is the tensor-product transformation of the transformations λ and μ.*

Proof. The mapping $(\lambda, \mu) \to \lambda \otimes_m \mu$ is bilinear. Hence there is a linear transformation σ: $U^* \otimes V^* \to (U \otimes V)^*$ such that $\sigma(\lambda \otimes \mu) = \lambda \otimes_m \mu$. Since $U^* \otimes V^*$ and $(U \otimes V)^*$ have the same dimension, it suffices to show that σ is onto.

Let $\{\alpha_1, \ldots, \alpha_m\}$ be a basis of U and $\{\beta_1, \ldots, \beta_n\}$ a basis of V. Let $\{\lambda_1, \ldots, \lambda_m\}$ be the dual basis to $\{\alpha_1, \ldots, \alpha_m\}$ and $\{\mu_1, \ldots, \mu_n\}$ the dual basis to $\{\beta_1, \ldots, \beta_n\}$. Then the elements $\lambda_1 \otimes_m \mu_1$, $\lambda_1 \otimes_m \mu_2, \ldots, \lambda_m \otimes_m \mu_n$ constitute a dual basis to the basis $\{\alpha_1 \otimes \beta_1, \alpha_1 \otimes \beta_2, \ldots, \alpha_m \otimes \beta_n\}$ of $U \otimes V$. It follows that σ maps $U^* \otimes V^*$ onto $(U \otimes V)^*$.

We see from this theorem that, if λ is from U^* and μ is from V^*, then $\lambda \otimes \mu$ may be interpreted either as an element of $U^* \otimes V^*$ or as the tensor-product transformation of the transformations λ and μ. We shall no longer distinguish between these points of view. Since the tensor multiplication of real numbers is the same as ordinary multiplication, we obtain the rule

$$(\lambda \otimes \mu)(\alpha \otimes \beta) = \lambda(\alpha)\mu(\beta)$$

EXERCISES

1. Let U, V, W, Z be finite-dimensional vector spaces. Prove that there is an isomorphism σ of $L(U; V) \otimes L(W; Z)$ onto $L(U \otimes W; V \otimes Z)$ such that $\sigma(\rho \otimes \tau)$ is the tensor-product transformation of the transformations ρ and τ.

2. Let σ_1: $U_1 \to V_1$ and σ_2: $U_2 \to V_2$ be linear transformations, where U_1, V_1, U_2, V_2 are finite-dimensional vector spaces. Prove that rank $(\sigma_1 \otimes \sigma_2) =$ (rank σ_1) (rank σ_2).

3. If σ: $U \to V$ is a linear transformation and U^* and V^* are the dual spaces to U and V, respectively, the linear transformation ${}^t\sigma$: $V^* \to W^*$ is defined by ${}^t\sigma(\lambda) = \lambda \circ \sigma$. If τ: $W \to Z$ is also a linear transformation, show that

$${}^t(\sigma \otimes \tau) = {}^t\sigma \otimes {}^t\tau$$

4. Let U, V, W, Z be finite-dimensional vector spaces with the bases $\{\alpha_1, \ldots, \alpha_m\}$, $\{\beta_1, \ldots, \beta_n\}$, $\{\gamma_1, \ldots, \gamma_r\}$, $\{\delta_1, \ldots, \delta_s\}$, respectively. Let σ: $U \to V$ and τ: $W \to Z$ be linear transformations with the matrices $\mathfrak{M}$ and $\mathfrak{N}$, respectively, with respect to the given bases. Find the matrix of $\sigma \otimes \tau$ with respect to the bases $\{\alpha_1 \otimes \gamma_1, \alpha_1 \otimes \gamma_2, \ldots, \alpha_1 \otimes \gamma_r, \alpha_2 \otimes \gamma_1, \ldots, \alpha_m \otimes \gamma_r\}$ and $\{\beta_1 \otimes \delta_1, \beta_1 \otimes \delta_2, \ldots, \beta_1 \otimes \delta_s, \beta_2 \otimes \delta_1, \ldots, \beta_n \otimes \delta_s\}$.

10-4 TENSOR SPACES

Let V be a vector space, and let V^* be its dual space. Denote by V_s^r the tensor product

$$V_s^r = \underbrace{V \otimes \cdots \otimes V}_{r \text{ factors}} \otimes \underbrace{V^* \otimes \cdots \otimes V^*}_{s \text{ factors}}$$

An element of V_s^r is called a *tensor of type* (r, s). r is called the *contravariant order*, and s the *covariant order* of such a tensor. V_0^0 is defined to be the real numbers. The elements of V_0^1 are called contravariant vectors, and the elements of V_1^0 are called covariant vectors.

In accordance with the commutative law for the tensor product, we shall disregard the order of the factors in V_s^r. For example, we shall also write

$$V_2^1 = V^* \otimes V \otimes V^*$$

if that is convenient.

Throughout the remainder of this section we shall denote tensor multiplication by juxtaposition. That is to say, if $\alpha_1, \ldots, \alpha_r$ are in V and $\lambda_1, \ldots, \lambda_s$ are in V^*, then $\alpha_1 \otimes \cdots \otimes \alpha_r \otimes \lambda_1 \otimes \cdots \otimes \lambda_s$ will be denoted by $\alpha_1\alpha_2 \cdots \alpha_r\lambda_1\lambda_2 \cdots \lambda_s$. Every element of V_s^r is a linear combination of elements of this type.

We shall now assume that V has finite dimension. Let $\{\alpha_1, \ldots, \alpha_n\}$ be a basis of V. Then there is a unique basis $\{\lambda_1, \ldots, \lambda_n\}$ of V^* which is dual to $\{\alpha_1, \ldots, \alpha_n\}$. Under these circumstances the n^{r+s} elements

$$\alpha_{i_1} \cdots \alpha_{i_r}\lambda_{j_1} \cdots \lambda_{j_s}$$

constitute a basis of V_s^r. If T is a tensor of type (r, s), we may express T uniquely in the form

$$\mathrm{T} = \sum_{i_1=1}^{n} \cdots \sum_{j_s=1}^{n} t_{j_1 \cdots j_s}^{i_1 \cdots i_r} \alpha_{i_1} \cdots \alpha_{i_r}\lambda_{j_1} \cdots \lambda_{j_s}$$

The coefficients $t_{j_1 \cdots j_s}^{i_1 \cdots i_r}$ are called the *components* of the tensor T with respect to the basis $\{\alpha_1, \ldots, \alpha_n\}$.

Let $\{\beta_1, \ldots, \beta_n\}$ be another basis of V. Then

$$\beta_i = \sum_{j=1}^{n} a_i{}^j\alpha_j$$

for $i = 1, \ldots, n$. If $\{\mu_1, \ldots, \mu_n\}$ is the dual basis to $\{\beta_1, \ldots, \beta_n\}$, then

$$\mu_i = \sum_{j=1}^{n} b_j{}^i \lambda_j$$

The product of the matrices $(b_j{}^i)$ and $(a_j{}^i)$ is then the identity matrix. We shall determine the effect of such a change of basis on the components of a tensor. Let $u_{j_1 \cdots j_s}^{i_1 \cdots i_r}$ be the components of the tensor T with respect to the basis $\{\beta_1, \ldots, \beta_n\}$. Then

$$\begin{aligned} \mathrm{T} &= \sum_{i_1=1}^{n} \cdots \sum_{j_s=1}^{n} u_{j_1 \cdots j_s}^{i_1 \cdots i_r} \beta_{i_1} \cdots \beta_{i_r} \mu_{j_1} \cdots \mu_{j_s} \\ &= \sum_{i_1=1}^{n} \cdots \sum_{j_s=1}^{n} u_{j_1 \cdots j_s}^{i_1 \cdots i_r} \sum_{k_1=1}^{n} \cdots \sum_{l_s=1}^{n} a_{i_1}{}^{k_1} \cdots a_{i_r}{}^{k_r} \\ &\qquad b_{l_1}{}^{j_1} \cdots b_{l_s}{}^{j_s} \alpha_{k_1} \cdots \alpha_{k_r} \lambda_{l_1} \cdots \lambda_{l} \end{aligned}$$

Hence if

$$\mathrm{T} = \sum_{k_1=1}^{n} \cdots \sum_{l_s=1}^{n} t_{l_1 \cdots l_s}^{k_1 \cdots k_r} \alpha_{k_1} \cdots \alpha_{k_r} \lambda_{l_1} \cdots \lambda_{l_s}$$

it follows that

$$t_{l_1 \cdots l_s}^{k_1 \cdots k_r} = \sum_{i_1=1}^{n} \cdots \sum_{j_s=1}^{n} u_{j_1 \cdots j_s}^{i_1 \cdots i_r} a_{i_1}{}^{k_1} \cdots a_{i_r}{}^{k_r} b_{l_1}{}^{j_1} \cdots b_{l_s}{}^{j_s}$$

This is the law of transformation of the components of a tensor of type (r, s).

Let $\sigma\colon V \to W$ be a linear transformation. The correspondence $\lambda \to \lambda \circ \sigma$ is then a linear transformation of W^* into V^* which is called the *transpose* of σ and is denoted by ${}^t\sigma$. If σ is an isomorphism of V onto W, then ${}^t\sigma$ will be an isomorphism of W^* onto V^*. In this case we may form the mapping ${}^t\sigma^{-1}$, which is an isomorphism of V^* onto W^*.

Now suppose that $\sigma\colon V \to V$ is an element of the general linear group on V. Then ${}^t\sigma^{-1}\colon V^* \to V^*$ is an element of the general linear group on V^*. We may form the tensor-product transformation

$$\sigma_s{}^r = \underbrace{\sigma \otimes \cdots \otimes \sigma}_{r \text{ factors}} \otimes \underbrace{{}^t\sigma^{-1} \otimes \cdots \otimes {}^t\sigma^{-1}}_{s \text{ factors}}$$

which will then be an element of the general linear group on $V_s{}^r$. It is essential to use the mapping ${}^t\sigma^{-1}$ on the covariant factors of $V_s{}^r$, instead of ${}^t\sigma$, so that the mapping will proceed in the same direction as it does for the contravariant factors.

Let $\theta: V_s^r \otimes V_{s'}^{r'} \to V_{s+s'}^{r+r'}$ be the natural isomorphism for which

$$\theta(\alpha_1 \cdots \alpha_r\lambda_1 \cdots \lambda_s \otimes \beta_1 \cdots \beta_{r'}\mu_1 \cdots \mu_{s'}) = \alpha_1 \cdots \alpha_r\beta_1 \cdots \beta_{r'}\lambda_1 \cdots \lambda_s\mu_1 \cdots \mu_{s'}$$

Then the bilinear mapping $\varphi: V_s^r \times V_{s'}^{r'} \to V_s^r \otimes V_{s'}^{r'}$, for which $\varphi(\mathrm{T}, \mathrm{T}') = \mathrm{T} \otimes \mathrm{T}'$, gives rise to a bilinear mapping $\theta \circ \varphi: V_s^r \times V_{s'}^{r'} \to V_{s+s'}^{r+r'}$ called the *multiplication of tensors*. If T, T′ are tensors from V_s^r, $V_{s'}^{r'}$, respectively, then the product TT′ is defined by

$$\mathrm{TT}' = (\theta \circ \varphi)(\mathrm{T}, \mathrm{T}')$$

In particular,

$$(\alpha_1 \cdots \alpha_r\lambda_1 \cdots \lambda_s)(\beta_1 \cdots \beta_{r'}\mu_1 \cdots \mu_{s'}) = \alpha_1 \cdots \alpha_r\beta_1 \cdots \beta_{r'}\lambda_1 \cdots \lambda_s\mu_1 \cdots \mu_{s'}$$

We shall now work out the product of tensors in terms of their components. Let $\{\alpha_1, \ldots, \alpha_n\}$ be a basis of V, and let $\{\lambda_1, \ldots, \lambda_n\}$ be the dual basis of V^*. Let

$$\mathrm{T} = \sum_{i_1=1}^{n} \cdots \sum_{j_s=1}^{n} t_{j_1 \cdots j_s}^{i_1 \cdots i_r} \alpha_{i_1} \cdots \alpha_{i_r}\lambda_{j_1} \cdots \lambda_{j_s}$$

and

$$\mathrm{T}' = \sum_{i_1=1}^{n} \cdots \sum_{j_{s'}=1}^{n} t'^{i_1 \cdots i_{r'}}_{j_1 \cdots j_{s'}} \alpha_{i_1} \cdots \alpha_{i_{r'}}\lambda_{j_1} \cdots \lambda_{j_{s'}}$$

be tensors from V_s^r and $V_{s'}^{r'}$, respectively. Then

$$\mathrm{TT}' = \sum_{i_1=1}^{n} \cdots \sum_{j_s=1}^{n} \sum_{k_1=1}^{n} \cdots \sum_{l_{s'}=1}^{n} t_{j_1 \cdots j_s}^{i_1 \cdots i_r} t'^{k_1 \cdots k_{r'}}_{l_1 \cdots l_{s'}} \alpha_{i_1} \cdots \alpha_{i_r}\alpha_{k_1} \cdots \alpha_{k_{r'}}\lambda_{j_1} \cdots \lambda_{j_s}\lambda_{l_1} \cdots \lambda_{l_{s'}}$$

Thus the components of TT′ are obtained by forming the products of the components of T and the components of T′.

EXERCISES

1. Prove that for $r \geqq 1$ and $s \geqq 1$ there is a linear transformation $\zeta_j^i: V_s^r \to V_{s-1}^{r-1}$ such that

$$\zeta_j^i(\alpha_1 \cdots \alpha_r\lambda_1 \cdots \lambda_s) = \lambda_j(\alpha_i)\alpha_1 \cdots \alpha_{i-1}\alpha_{i+1} \cdots \alpha_r\lambda_1 \cdots \lambda_{j-1}\lambda_{j+1} \cdots \lambda_s$$

Prove that, if σ is an element of the general linear group on V, then

$$\zeta_j^i \circ \sigma_s^r = \sigma_{s-1}^{r-1} \circ \zeta_j^i$$

2. Let τ be the natural isomorphism of $L(V; V)$ onto V_1^1. If ζ_1^1 is the mapping of Exercise 1 and σ is an element of $L(V; V)$, the *trace* of σ is defined by $\mathrm{Tr}(\sigma) = (\zeta_1^1 \circ \tau)(\sigma)$. Since Tr: $L(V; V) \to$ real numbers is a linear transformation, it is a linear functional of $L(V; V)$. Let $\mathfrak{M}$ be the matrix of σ with respect to a basis $\{\alpha_1, \ldots, \alpha_n\}$ of V. Find Tr (σ) in terms of the components of $\mathfrak{M}$. Prove that

$$\operatorname{Tr}(\sigma_1\sigma_2) = \operatorname{Tr}(\sigma_2\sigma_1)$$

10-5 SYMMETRY AND ALTERNATION

Let S_r denote the symmetric group on r symbols, that is, the group of all permutations of the symbols $\{1, \ldots, r\}$. Through the commutative law for tensor multiplication, the elements of S_r give rise naturally to linear mappings of $V_0{}^r$ (or of $V_r{}^0$) onto itself; namely, if π is an element of S_r, there is an isomorphism σ of $V_0{}^r$ onto $V_0{}^r$ such that

$$\sigma(\alpha_1 \otimes \cdots \otimes \alpha_r) = \alpha_{\pi(1)} \otimes \cdots \otimes \alpha_{\pi(r)}$$

(and similarly for $V_r{}^0$). We shall denote this mapping by π also.

A tensor T in $V_0{}^r$ is called *symmetric* if $\pi(\mathrm{T}) = \mathrm{T}$ for all π in S_r. It is called *alternating* if $\pi(\mathrm{T}) = (\operatorname{sign} \pi)\mathrm{T}$ for all π in S_r. A linear transformation $\sigma\colon V_0{}^r \to W$ is called *symmetric* if $\sigma\circ\pi = \sigma$ for all π in S_r. It is called *alternating* if $\sigma\circ\pi = (\operatorname{sign} \pi)\sigma$ for all π in S_r.

We have seen that the tensor space $V_r{}^0$ is naturally isomorphic to $(V_0{}^r)^*$. If $\lambda_1, \ldots, \lambda_r$ are elements of V^*, we may regard $\lambda_1 \otimes \cdots \otimes \lambda_r$ either as an element of $V_r{}^0$ or as the tensor-product transformation of the transformations $\lambda_1, \ldots, \lambda_r$. The definitions of symmetry and alternation are not the same from these two points of view, but we shall see that they are equivalent.

Theorem 10-10. *Let $\sigma\colon V_r{}^0 \to (V_0{}^r)^*$ be the natural isomorphism. If* T^* *is a tensor from $V_r{}^0$, then*

$$\sigma(\pi(\mathrm{T}^*)) = \sigma(\mathrm{T}^*)\circ\pi^{-1}$$

for all π in S_r.

Proof. Denote by $\otimes_m$ the tensor multiplication of transformations. Let $\lambda_1, \ldots, \lambda_r$ be elements of V^* and $\alpha_1, \ldots, \alpha_r$ be elements of V. Then

$$\begin{aligned}\sigma[\pi(\lambda_1 \otimes \cdots \otimes \lambda_r)](\alpha_1 \otimes \cdots \otimes \alpha_r) &\\ = (\lambda_{\pi(1)} \otimes_m \cdots \otimes_m \lambda_{\pi(r)})(\alpha_1 \otimes \cdots \otimes \alpha_r) &\\ = \lambda_{\pi(1)}(\alpha_1) \cdots \lambda_{\pi(r)}(\alpha_r) &\end{aligned}$$

and

$$\begin{aligned}[\sigma(\lambda_1 \otimes \cdots \otimes \lambda_r)\circ\pi^{-1}](\alpha_1 \otimes \cdots \otimes \alpha_r) &\\ = (\lambda_1 \otimes_m \cdots \otimes_m \lambda_r)(\alpha_{\pi^{-1}(1)} \otimes \cdots \otimes \alpha_{\pi^{-1}(r)}) &\\ = \lambda_1(\alpha_{\pi^{-1}(1)}) \cdots \lambda_r(\alpha_{\pi^{-1}(r)}) &\\ = \lambda_{\pi(1)}(\alpha_1) \cdots \lambda_{\pi(r)}(\alpha_r) &\end{aligned}$$

Hence the theorem is true for $T^* = \lambda_1 \otimes \cdots \otimes \lambda_r$ and so for all elements of V_r^0.

It follows immediately from this theorem that a tensor T^* of V_r^0 is symmetric if and only if it is symmetric as a linear functional of V_0^r, and likewise for alternation.

The linear transformation

$$\Sigma_r = \sum_{S_r} \pi$$

of V_0^r into itself is called *symmetrization*. The linear transformation

$$A_r = \sum_{S_r} (\text{sign } \pi)\pi$$

of V_0^r into itself is called *alternation*.

Theorem 10-11. *Let* T *be a tensor in* V_0^r. *Then* $\Sigma_r(T)$ *is symmetric and* $A_r(T)$ *is alternating. Furthermore, if* T *is a symmetric tensor, then*

$$\Sigma_r(T) = r!T$$

and if T *is alternating, then*

$$A_r(T) = r!T$$

Proof. Let π_0 be an element of S_r. Then

$$\pi_0(\Sigma_r(T)) = \sum_{S_r} \pi_0\pi(T)$$

$$\pi_0(A_r(T)) = \sum_{S_r} (\text{sign } \pi)\pi_0\pi(T)$$

As π runs through the elements of S_r, so does $\pi_0\pi$. Furthermore, sign π = (sign π_0)(sign $\pi_0\pi$). Hence $\pi_0(\Sigma_r(T)) = \Sigma_r(T)$ and

$$\pi_0(A_r(T)) = (\text{sign } \pi_0)A_r(T))$$

This proves the symmetry of $\Sigma_r(T)$ and the alternation of $A_r(T)$.

If T is symmetric, then

$$\Sigma_r(T) = \sum_{S_r} \pi(T) = \sum_{S_r} T = r!T$$

If T is alternating, then

$$A_r(T) = \sum_{S_r} (\text{sign } \pi)\pi(T) = \sum_{S_r} (\text{sign } \pi)^2\, T = \sum_{S_r} T = r!T$$

It follows from this theorem that Σ_r is a linear transformation of V_0^r onto the subspace of V_0^r consisting of the symmetric tensors. Furthermore, A_r is a linear transformation of V_0^r onto the subspace of V_0^r consisting of the alternating tensors.

Theorem 10-12. *Let N^r denote the kernel of the alternation mapping* A_r. *Let* $\sigma\colon V_0{}^r \to W$ *be a linear transformation. Then* σ *is alternating if and only if* σ *carries the elements of* N^r *onto zero.*

Proof. Let σ be alternating. Let T be an element of N^r. Then $A_r(T) = 0$, and so $\sigma(A_r(T)) = 0$. But

$$\sigma(A_r(T)) = \sigma\left(\sum_{S_r} (\text{sign } \pi)\pi(T)\right) = \sum_{S_r} (\text{sign } \pi)\sigma(\pi(T))$$
$$= \sum_{S_r} \sigma(T) = r!\sigma(T)$$

Hence $\sigma(T) = 0$.

Conversely, let σ carry the elements of N^r onto zero. If T is in $V_0{}^r$ and π_0 is in S_r, then

$$A_r[T - (\text{sign } \pi_0)\pi_0(T)] = \sum_{S_r} (\text{sign } \pi)[\pi(T) - (\text{sign } \pi_0)\pi\pi_0(T)]$$
$$= \sum_{S_r} (\text{sign } \pi)\pi(T) - \sum_{S_r} (\text{sign } \pi\pi_0)\pi\pi_0(T) = 0$$

that is, $T - (\text{sign } \pi_0)\pi_0(T)$ is in N^r. Thus

$$\sigma(T) = (\text{sign } \pi_0)\sigma(\pi_0(T))$$

and so $\sigma \circ \pi_0 = (\text{sign } \pi_0)\sigma$, which shows that σ is alternating.

EXERCISES

1. Let Z^r be the kernel of the symmetry mapping Σ_r. Prove that a linear transformation $\sigma\colon V_0{}^r \to W$ is symmetric if and only if σ carries the elements of Z^r onto zero.

2. Let Q be the subspace of $V_0{}^r$ spanned by all elements of the form $\alpha_1 \otimes \cdots \otimes \alpha_r$ with two or more factors the same. Prove that $Q \subset N^r$.

3. Let V be a vector space of finite dimension with the basis $\{\alpha_1, \ldots, \alpha_n\}$. Then the elements $B = \{\alpha_{i_1} \otimes \cdots \otimes \alpha_{i_r}\}$ with $1 \leqq i_j \leqq n$ for $j = 1, \ldots, r$ constitute a basis for $V_0{}^r$. Write $B = B_0 \cup B_1$, where B_0 consists of those elements with two or more factors the same and $B_0 \cup B_1$ is empty. If $\alpha_{i_1} \otimes \cdots \otimes \alpha_{i_r}$ and $\alpha_{j_1} \otimes \cdots \otimes \alpha_{j_r}$ are from B_1, we say that they are *equivalent* when there is a permutation π from S_r such that $i_k = j_{\pi(k)}$ for $k = 1, \ldots, r$. By means of this relation, B_1 may be divided into equivalence classes, each of which is made up of tensor products which differ only in the order of their factors. Choose one element from each equivalence class and denote by C_1 the set of these selected elements. Denote by S_r' the set of elements of S_r which are different from the identity permutation. Then

$$B = B_0 \cup C_1 \bigcup_{S_r'} \pi(C_1)$$

Prove that, if

$$T = \sum_{C_1} a_{i_1 \cdots i_r} \alpha_{i_1} \otimes \cdots \otimes \alpha_{i_r}$$

and $A_r(T) = 0$, then $T = 0$.

4. Using the notation of Exercise 3, let C_π consist of the elements

$$\alpha_{i_1} \otimes \cdots \otimes \alpha_{i_r} - (\text{sign } \pi)\pi(\alpha_{i_1} \otimes \cdots \otimes \alpha_{i_r})$$

where $\alpha_{i_1} \otimes \cdots \otimes \alpha_{i_r}$ is in C_1. Prove that

$$B_0 \cup C_1 \bigcup_{S_r'} C_\pi$$

is also a basis for $V_0{}^r$. Using this result and the preceding exercises, prove that $Q = N^r$.

10-6 EXTERIOR POWERS OF A VECTOR SPACE

Let V be a vector space. Let N^r be the kernel of the alternation mapping A_r: $V_0{}^r \to V_0{}^r$. The quotient space $V_0{}^r/N^r$ is called the

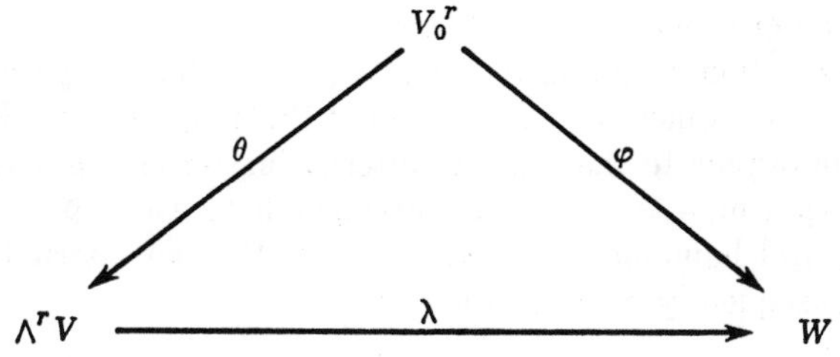

FIG. 10-3

r-fold exterior power of V. It is denoted by $\wedge^r V$. Then $\wedge^1 V = V$ and we define $\wedge^0 V$ = real numbers.

The next theorem shows that the r-fold exterior power plays the same role with respect to alternating mappings that the tensor product plays with respect to multilinear mappings (see Theorem 10-1).

Theorem 10-13. *If φ: $V_0{}^r \to W$ is an alternating linear transformation, then there is a uniquely determined linear transformation λ: $\wedge^r V \to W$ such that $\varphi = \lambda \circ \theta$, where θ: $V_0{}^r \to \wedge^r V$ is the natural homomorphism. Furthermore, the mapping $\varphi \to \lambda$ is an isomorphism of the vector space $A(V_0{}^r; W)$ of all alternating linear transformations of $V_0{}^r$ into W onto the vector space $L(\wedge^r V; W)$.*

Proof. We have seen that φ carries the elements of N^r onto zero. Hence we may define

$$\lambda(\xi + N^r) = \varphi(\xi)$$

where $\xi + N^r$ is an element of $\wedge^r V$. Since $\theta(\xi) = \xi + N^r$, it follows immediately that $\varphi = \lambda \circ \theta$, and it is clear that λ is unique (see Figure 10-3).

The mapping $\varphi \to \lambda$ is obviously an isomorphism. We must show that it is onto $L(\wedge^r V; W)$. Let λ be an element of $L(\wedge^r V; W)$. Define $\varphi = \lambda \circ \theta$. Then φ carries the elements of N^r onto zero, and we have seen that this implies that φ is alternating. Since $\varphi \to \lambda$, the proof is complete.

If $\alpha_1, \ldots, \alpha_r$ are elements of V, the element

$$\alpha_1 \otimes \cdots \otimes \alpha_r + N^r$$

of $\wedge^r V$ is denoted by

$$\alpha_1 \wedge \alpha_2 \wedge \cdots \wedge \alpha_r$$

Such an element of $\wedge^r V$ is said to be *decomposable.* Since every element of $V_0{}^r$ is a linear combination of elements of the form $\alpha_1 \otimes \cdots \otimes \alpha_r$, it follows that every element of $\wedge^r V$ is a linear combination of elements of the form $\alpha_1 \wedge \cdots \wedge \alpha_r$.

The alternation mapping A_r maps $V_0{}^r$ onto the subspace of alternating tensors. Since N^r is the kernel of this mapping, it follows that $\wedge^r V$ is isomorphic to the space of alternating tensors in $V_0{}^r$. Under this isomorphism, $\alpha_1 \wedge \cdots \wedge \alpha_r$ corresponds to $A_r(\alpha_1 \otimes \cdots \otimes \alpha_r)$.

The natural homomorphism θ: $V_0{}^r \to \wedge^r V$ is an alternating mapping. Consequently we have the rule

$$\alpha_{\pi(1)} \wedge \cdots \wedge \alpha_{\pi(r)} = (\text{sign } \pi)\alpha_1 \wedge \cdots \wedge \alpha_r$$

for π in S_r.

Theorem 10-14. *Let σ: $V \to W$ be a linear transformation. Then there exists a linear transformation $\wedge^r \sigma$: $\wedge^r V \to \wedge^r W$ such that*

$$\wedge^r \sigma(\alpha_1 \wedge \cdots \wedge \alpha_r) = \sigma(\alpha_1) \wedge \cdots \wedge \sigma(\alpha_r)$$

Proof. Define τ: $V_0{}^r \to W_0{}^r$ by

$$\tau = \sigma \otimes \cdots \otimes \sigma \qquad (r \text{ factors})$$

Let η: $W_0{}^r \to \wedge^r W$ be the natural homomorphism. Since η is alternating and $\pi \circ \tau = \tau \circ \pi$,

$$\eta \circ \tau \circ \pi = \eta \circ \pi \circ \tau = (\text{sign } \pi)\eta \circ \tau$$

which proves that $\eta \circ \tau$ is alternating. Hence there is a linear transformation $\wedge^r \sigma$: $\wedge^r V \to \wedge^r W$ such that $\eta \circ \tau = (\wedge^r \sigma) \circ \theta$, where θ: $V_0{}^r \to \wedge^r V$ is the natural homomorphism (see Figure 10-4). If $\alpha_1, \ldots, \alpha_r$ are elements of V, then

$$\begin{aligned} \eta \circ \tau(\alpha_1 \otimes \cdots \otimes \alpha_r) &= \eta(\sigma(\alpha_1) \otimes \cdots \otimes \sigma(\alpha_r)) \\ &= \sigma(\alpha_1) \wedge \cdots \wedge \sigma(\alpha_r) \end{aligned}$$

and

$$[(\wedge^r\sigma)\circ\theta](\alpha_1 \otimes \cdots \otimes \alpha_r) = \wedge^r\sigma(\alpha_1 \wedge \cdots \wedge \alpha_r)$$

so that the theorem is proved.

The mapping $\wedge^r\sigma$ is called the *r-fold exterior power* of σ.

Let $\sigma: V \to W$ be a linear transformation and ${}^t\sigma: W^* \to V^*$ its transpose. Then the transpose of $\wedge^r\sigma$ is $\wedge^r\, {}^t\sigma$.

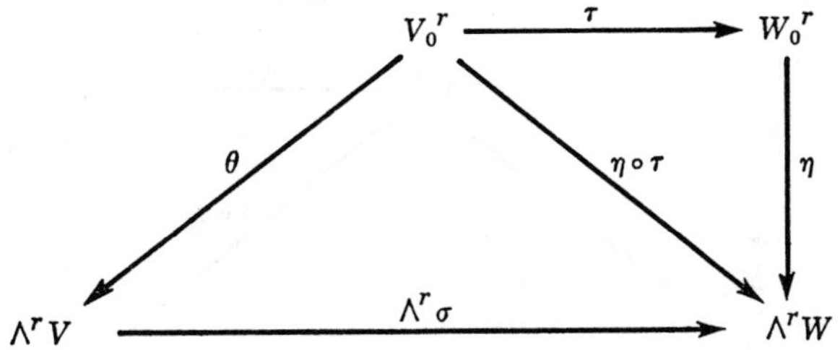

FIG. 10-4

EXERCISES

1. Prove the last statement of Section 10-6.

2. Let V be a vector space. Let Z^r be the kernel of the symmetry mapping $\Sigma_r: V_0^r \to V_0^r$. Define $\dagger^r V = V_0^r/Z^r$. Let $\theta: V_0^r \to \dagger^r V$ be the natural homomorphism. Prove that if $\varphi: V_0^r \to W$ is a symmetric linear transformation there is a uniquely determined linear transformation $\lambda: \dagger^r V \to W$ such that $\varphi = \lambda\circ\theta$. Prove that the mapping $\varphi \to \lambda$ is an isomorphism of the vector space $S(V_0^r; W)$ of all symmetric linear transformations of V_0^r into W onto $L(\dagger^r V; W)$.

3. Using the notation of Exercise 2, denote the element $\alpha_1 \otimes \cdots \otimes \alpha_r + Z^r$ of $\dagger^r V$ by $\alpha_1 \dagger \cdots \dagger \alpha_r$, Prove that if π is a permutation from S_r then $\alpha_{\pi(1)} \dagger \cdots \dagger \alpha_{\pi(r)} = \alpha_1 \dagger \cdots \dagger \alpha_r$.

4. Using the notation of Exercises 2 and 3, prove that if $\sigma: V \to W$ is a linear transformation, there exists a linear transformation $\dagger^r\sigma: \dagger^r V \to \dagger^r W$ such that

$$\dagger^r\sigma(\alpha_1 \dagger \cdots \dagger \alpha_r) = \sigma(\alpha_1) \dagger \cdots \dagger \sigma(\alpha_r)$$

5. Using the notation of the preceding exercises, prove that, if $\sigma: V \to W$ is a linear transformation and if ${}^t\sigma: W^* \to V^*$ is the transpose of σ, the transpose of $\dagger^r\sigma$ is $\dagger^r\, {}^t\sigma$.

6. If V = real numbers, what are the spaces $\wedge^r V$ and $\dagger^r V$?

10-7 EXTERIOR ALGEBRA

Theorem 10-15. *There is a bilinear mapping* $\mu: \wedge^r V \times \wedge^s V \to \wedge^{r+s} V$ *such that*

$$\mu(\alpha_1 \wedge \cdots \wedge \alpha_r, \beta_1 \wedge \cdots \wedge \beta_s) = \alpha_1 \wedge \cdots \wedge \alpha_r \wedge \beta_1 \wedge \cdots \wedge \beta_s$$

Proof. For each tensor T of V_0^s the mapping

$$(\alpha_1, \ldots, \alpha_r) \to \alpha_1 \otimes \cdots \otimes \alpha_r \otimes \mathrm{T}$$

is multilinear and so determines a linear transformation $\sigma_\mathrm{T}\colon V_0^r \to V_0^{r+s}$ such that

$$\sigma_\mathrm{T}(\alpha_1 \otimes \cdots \otimes \alpha_r) = \alpha_1 \otimes \cdots \otimes \alpha_r \otimes \mathrm{T}$$

Let $\nu\colon V_0^{r+s} \to \wedge^{r+s}V$ be the natural homomorphism. Then $\nu \circ \sigma_\mathrm{T}\colon V_0^r \to \wedge^{r+s}V$ is an alternating linear transformation and so determines

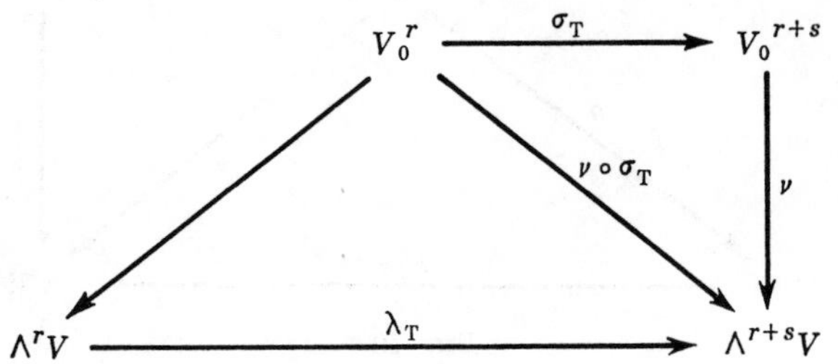

FIG. 10-5

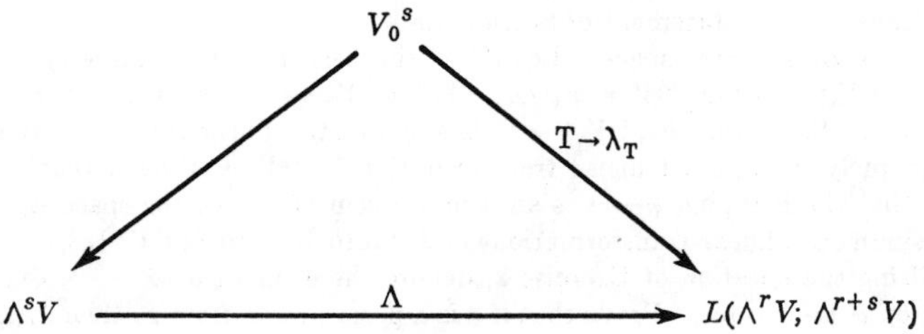

FIG. 10-6

a linear transformation $\lambda_\mathrm{T}\colon \wedge^r V \to \wedge^{r+s}V$ such that, if

$$\mathrm{T} = \beta_1 \otimes \cdots \otimes \beta_s$$

then

$$\lambda_\mathrm{T}(\alpha_1 \wedge \cdots \wedge \alpha_r) = \alpha_1 \wedge \cdots \wedge \alpha_r \wedge \beta_1 \wedge \cdots \wedge \beta_s$$

(see Figure 10-5).

The mapping $\mathrm{T} \to \lambda_\mathrm{T}$ of V_0^s into $L(\wedge^r V; \wedge^{r+s}V)$ is then an alternating linear transformation and so determines a linear transformation $\Lambda\colon \wedge^s V \to L(\wedge^r V; \wedge^{r+s}V)$ such that, if $\lambda = \Lambda(\beta_1 \wedge \cdots \wedge \beta_s)$, then

$$\lambda(\alpha_1 \wedge \cdots \wedge \alpha_r) = \alpha_1 \wedge \cdots \wedge \alpha_r \wedge \beta_1 \wedge \cdots \wedge \beta_s$$

(Figure 10-6).

Let ζ be an element of $\wedge^r V$ and η an element of $\wedge^s V$. Define

$$\mu(\zeta, \eta) = \lambda(\zeta)$$

where $\lambda = \Lambda(\eta)$. Then $\mu: \wedge^r V \times \wedge^s V \to \wedge^{r+s} V$ is clearly bilinear and satisfies the desired condition. This completes the proof.

If ζ is an element of $\wedge^r V$ and η is an element of $\wedge^s V$, then $\mu(\zeta, \eta)$ will be denoted by $\zeta \wedge \eta$. This element of $\wedge^{r+s} V$ is called the *exterior product* of ζ and η. We have seen that it is bilinear and satisfies

$$(\alpha_1 \wedge \cdots \wedge \alpha_r) \wedge (\beta_1 \wedge \cdots \wedge \beta_s) = \alpha_1 \wedge \cdots \wedge \alpha_r \wedge \beta_1 \wedge \cdots \wedge \beta_s$$

The exterior product therefore satisfies the associative law

$$[(\alpha_1 \wedge \cdots \wedge \alpha_r) \wedge (\beta_1 \wedge \cdots \wedge \beta_s)] \wedge (\gamma_1 \wedge \cdots \wedge \gamma_t)$$
$$= (\alpha_1 \wedge \cdots \wedge \alpha_r) \wedge [(\beta_1 \wedge \cdots \wedge \beta_s) \wedge (\gamma_1 \wedge \cdots \wedge \gamma_t)]$$

for decomposable elements. By forming linear combinations of these, we may prove the associative law in general:

$$(\zeta \wedge \eta) \wedge \xi = \zeta \wedge (\eta \wedge \xi)$$

where ζ is in $\wedge^r V$, η is in $\wedge^s V$, and ξ is in $\wedge^t V$.

Theorem 10-16. *If ζ is in $\wedge^r V$ and η is in $\wedge^s V$, then*

$$\eta \wedge \zeta = (-1)^{rs} \zeta \wedge \eta$$

Proof. Let $\zeta = \alpha_1 \wedge \cdots \wedge \alpha_r$ and $\eta = \beta_1 \wedge \cdots \wedge \beta_s$. Let π be the permutation in S_{r+s} such that

$$\pi(\alpha_1 \otimes \cdots \otimes \alpha_r \otimes \beta_1 \otimes \cdots \otimes \beta_s)$$
$$= \beta_1 \otimes \cdots \otimes \beta_s \otimes \alpha_1 \otimes \cdots \otimes \alpha_r$$

Then sign $\pi = (-1)^{rs}$ and so

$$\beta_1 \wedge \cdots \wedge \beta_s \wedge \alpha_1 \wedge \cdots \wedge \alpha_r$$
$$= (-1)^{rs} \alpha_1 \wedge \cdots \wedge \alpha_r \wedge \beta_1 \wedge \cdots \wedge \beta_s$$

The theorem then follows by forming linear combinations of decomposable elements.

Theorem 10-17. *If V has dimension n and if $r > n$, then $\wedge^r V$ is the zero space.*

Proof. Let $\{\alpha_1, \ldots, \alpha_n\}$ be a basis of V. Every element of $\wedge^r V$ is a linear combination of elements of the form $\alpha_{i_1} \wedge \cdots \wedge \alpha_{i_r}$. But such a decomposable element has at least two of its vectors the same. If we interchange two equal vectors, the element must be multiplied by -1 and yet remain unchanged. Hence it must be zero. It follows that every element of $\wedge^r V$ is zero.

Let V be a vector space of dimension n. Denote by $\wedge V$ the direct sum of the vector spaces $\wedge^0 V, \wedge^1 V, \ldots, \wedge^n V$. An element Ξ of $\wedge V$ may be written uniquely as

$$\Xi = \xi_0 + \xi_1 + \cdots + \xi_n$$

where ξ_i is an element of $\wedge^i V$. If

$$\Xi' = \xi_0' + \xi_1' + \cdots + \xi_n'$$

is another element of $\wedge V$, we may define

$$\Xi\Xi' = \eta_0 + \eta_1 + \cdots + \eta_n$$

where
$$\eta_i = \sum_{j=0}^{i} \xi_j \wedge \xi_{i-j}'$$

and where $\xi_0 \wedge \xi_i'$ is defined to be $\xi_0 \xi_i'$ and $\xi_i \wedge \xi_0'$ is defined to be $\xi_0' \xi_i$. Then $\wedge V$ is an algebra (a vector space which is also a ring) called the *exterior algebra* (*Grassman algebra*) of V.

Theorem 10-18. *Let $\{\alpha_1, \ldots, \alpha_n\}$ be a basis of the vector space V. Then the $\binom{n}{r}$ elements*

$$\alpha_{i_1} \wedge \cdots \wedge \alpha_{i_r} \qquad 1 \leqq i_1 < i_2 < \cdots < i_r \leqq n$$
$$r = 1, \ldots, n$$

form a basis of the exterior power $\wedge^r V$. The exterior algebra $\wedge V$ therefore has the dimension 2^n.

Proof. It is clear from the preceding theorem that the elements

$$\alpha_{i_1} \wedge \cdots \wedge \alpha_{i_r} \qquad 1 \leqq i_1 < \cdots < i_r \leqq n$$

span $\wedge^r V$. It suffices to prove that these elements are linearly independent. Suppose that

$$\Sigma a_{i_1 \ldots i_r} \alpha_{i_1} \wedge \cdots \wedge \alpha_{i_r} = 0$$

Let $j_1, \ldots, j_{n-r}$ be $n - r$ integers from the set $\{1, \ldots, n\}$. Multiply the relation by $\alpha_{j_1} \wedge \cdots \wedge \alpha_{j_{n-r}}$. Then

$$a_{k_1 \ldots k_r} \alpha_{k_1} \wedge \cdots \wedge \alpha_{k_r} \wedge \alpha_{j_1} \wedge \cdots \wedge \alpha_{j_{n-r}} = 0$$

where $k_1, \ldots, k_r$ are those integers such that

$$\{k_1, \ldots, k_r, j_1, \ldots, j_{n-r}\}$$

is the set $\{1, \ldots, n\}$. All other terms of the sum are zero because of repeated factors.

If we can prove that $\alpha_1 \wedge \cdots \wedge \alpha_n \neq 0$, then

$$\alpha_{k_1} \wedge \cdots \wedge \alpha_{k_r} \wedge \alpha_{j_1} \wedge \cdots \wedge \alpha_{j_{n-r}} \neq 0$$

and we may conclude that $a_{k_1 \ldots k_r} = 0$. If we apply the alternation mapping A_n to the tensor $\alpha_1 \otimes \cdots \otimes \alpha_n$, we do not obtain the zero element, for the elements $\{\pi(\alpha_1 \otimes \cdots \otimes \alpha_n)\}$ with π in S_r are linearly independent. Hence $\alpha_1 \wedge \cdots \wedge \alpha_n \neq 0$.

In this way we may prove that each coefficient $a_{i_1 \ldots i_r}$ is zero and thus conclude the proof.

On the basis of this theorem, we may write each element ξ of $\wedge^r V$ uniquely in the form

$$\xi = \sum_{1 \leqq i_1 < \cdots < i_r \leqq n} a_{i_1 \ldots i_r} \alpha_{i_1} \wedge \cdots \wedge \alpha_{i_r}$$

On occasion it is more convenient to write

$$\xi = \sum_{i_1=1}^{n} \cdots \sum_{i_r=1}^{n} b_{i_1 \ldots i_r} \alpha_{i_1} \wedge \cdots \wedge \alpha_{i_r}$$

The coefficients $b_{i_1 \ldots i_r}$ are uniquely determined if we impose the condition

$$b_{i_{\pi(1)} \ldots i_{\pi(r)}} = (\text{sign } \pi) b_{i_1 \ldots i_r}$$

for all permutations π in S_r. They have the values

$$b_{j_1 \ldots j_r} = \begin{cases} 0 & \text{if two or more of the integers } j_1, \ldots, \\ & j_r \text{ are equal} \\ \dfrac{\text{sign } \pi}{r!} a_{j_{\pi(1)} \cdots j_{\pi(r)}} & \text{where } \pi \text{ is the permutation for which} \\ & j_{\pi(1)} < j_{\pi(2)} < \cdots < j_{\pi(r)} \end{cases}$$

and are called the *components* of ξ.

Let $\{\alpha_1, \ldots, \alpha_n\}$ be a basis of V. Let $\beta_1, \ldots, \beta_r$ be vectors of V. Then

$$\beta_i = b_{i1}\alpha_1 + \cdots + b_{in}\alpha_n$$

for $i = 1, \ldots, r$. We shall determine the coefficients of $\beta_1 \wedge \cdots \wedge \beta_r$ in terms of the $r \times n$ matrix (b_{ij}). On the one hand,

$$\beta_1 \wedge \cdots \wedge \beta_r = \sum_{i_1=1}^{n} \cdots \sum_{i_r=1}^{n} b_{1i_1} \ldots b_{ri_r} \alpha_{i_1} \wedge \cdots \wedge \alpha_{i_r}$$

and, on the other hand,

$$\beta_1 \wedge \cdots \wedge \beta_r = \sum_{1 \leqq i_1 < \cdots < i_r \leqq n} a_{i_1 \ldots i_r} \alpha_{i_1} \wedge \cdots \wedge \alpha_{i_r}$$

Hence

$$a_{i_1 \ldots i_r} = \sum_{S_r} (\text{sign } \pi) b_{1 i_{\pi(1)}} \ldots b_{r i_{\pi(r)}}$$

We see that the coefficient $a_{i_1 \ldots i_r}$ is the determinant of the $r \times r$ minor of (b_{ij}) consisting of the columns numbered $i_1, \ldots, i_r$ In particular, if $r = n$, then

$$\beta_1 \wedge \cdots \wedge \beta_n = \det(b_{ij})\, \alpha_1 \wedge \cdots \wedge \alpha_n$$

Theorem 10-19. *Let $\beta_1, \ldots, \beta_r$ be vectors from a finite-dimensional vector space V. A necessary and sufficient condition that these vectors be linearly dependent is that*

$$\beta_1 \wedge \cdots \wedge \beta_r = 0$$

Proof. If $\beta_1, \ldots, \beta_r$ are linearly independent, they may be made part of a basis of V. Then $\beta_1 \wedge \cdots \wedge \beta_r$ is part of a basis of $\wedge^r V$ and so is not equal to zero.

Conversely, if $\beta_1, \ldots, \beta_r$ are linearly dependent, say $\Sigma b_i \beta_i = 0$ with $b_1 \neq 0$, multiply by $\beta_2 \wedge \cdots \wedge \beta_r$. Then $b_1 \beta_1 \wedge \cdots \wedge \beta_r = 0$, and so $\beta_1 \wedge \cdots \wedge \beta_r = 0$.

By means of this theorem it is possible to describe the subspaces of a finite-dimensional vector space in terms of the exterior algebra of V. Let $\alpha_1, \ldots, \alpha_r$ be linearly independent vectors of V. Then a vector β from V is a linear combination of $\alpha_1, \ldots, \alpha_r$ if and only if $\beta \wedge \alpha_1 \wedge \cdots \wedge \alpha_r = 0$. We may therefore conclude that *the set W of those vectors β such that $\beta \wedge \alpha_1 \wedge \cdots \wedge \alpha_r = 0$ is precisely the subspace of V which has the basis $\{\alpha_1, \ldots, \alpha_r\}$.* If $\{\alpha_1, \ldots, \alpha_r\}$ and $\{\alpha'_1, \ldots, \alpha'_r\}$ are two bases of W, then

$$\alpha'_1 \wedge \cdots \wedge \alpha'_r = \det(a_{ij})\alpha_1 \wedge \cdots \wedge \alpha_r$$

where (a_{ij}) is the matrix of the relations which express $\alpha'_1, \ldots, \alpha'_r$ in terms of $\alpha_1, \ldots, \alpha_r$. Thus the element $\alpha_1 \wedge \cdots \wedge \alpha_r$ of $\wedge^r V$ which is associated with W is uniquely determined except for a numerical factor.

EXERCISES

The notation of the previous set of exercises will be employed in the following ones.

1. Prove that there is a bilinear mapping $\mu\colon \dagger^r V \times \dagger^s V \to \dagger^{r+s} V$ such that

$$\mu(\alpha_1 \dagger \cdots \dagger \alpha_r, \beta_1 \dagger \cdots \dagger \beta_s) = \alpha_1 \dagger \cdots \dagger \alpha_r \dagger \beta_1 \dagger \cdots \dagger \beta_s$$

2. Using the mapping μ of Exercise 1, define $\xi \dagger \eta = \mu(\xi, \eta)$, where ξ is an element of $\dagger^r V$ and η is an element of $\dagger^s V$. Prove that this product is associative and commutative.

3. If $\{\alpha_1, \ldots, \alpha_n\}$ is a basis of V, prove that the elements

$$\alpha_{i_1} \dagger \cdots \dagger \alpha_{i_r} \qquad 1 \leqq i_1 \leqq \cdots \leqq i_r \leqq n$$

constitute a basis of $\dagger^r V$.

4. Let U be a finite-dimensional vector space. Let V and W be subspaces of V with the bases $\{\alpha_1, \ldots, \alpha_m\}$ and $\{\beta_1, \ldots, \beta_n\}$, respectively. Let $\xi = \alpha_1 \wedge \cdots \wedge \alpha_m$ and $\eta = \beta_1 \wedge \cdots \wedge \beta_n$.

a. Prove that $V \subset W$ if and only if there are vectors $\gamma_1, \ldots, \gamma_{n-m}$ in U such that $\xi \wedge \gamma_1 \wedge \cdots \wedge \gamma_{n-m} = \eta$.

b. Prove that $V \cap W = \{0\}$ if and only if $\xi \wedge \eta \neq 0$.

10-8 DUALITY IN EXTERIOR ALGEBRA

Let V be a finite-dimensional vector space, and let V^* be the dual space of V. We have seen how it is possible to identify the vector space $V_r^0 = V^* \otimes \cdots \otimes V^*$ with the vector space

$$(V_0^r)^* = (V \otimes \cdots \otimes V)^*$$

Indeed, if $\lambda_1, \ldots, \lambda_r$ are elements of V^*, then $\lambda_1 \otimes \cdots \otimes \lambda_r$ may be regarded either as a tensor in V_r^0 or as the tensor-product transformation of the transformations $\lambda_1, \ldots, \lambda_r$ and thus an element of $(V_0^r)^*$. The following theorem presents the analogous result for the exterior power $\wedge^r V$.

Theorem 10-20. *Let V be a finite-dimensional vector space and V^* its dual space. Then there is an isomorphism σ of $\wedge^r(V^*)$ onto $(\wedge^r V)^*$ such that*

$$[\sigma(\lambda_1 \wedge \cdots \wedge \lambda_r)](\alpha_1 \wedge \cdots \wedge \alpha_r) = \det(\lambda_i(\alpha_j))$$

Proof. Let $\tau\colon V_r^0 \to (V_0^r)^*$ be the natural isomorphism under which $\tau(\lambda_1 \otimes \cdots \otimes \lambda_r)$ is the tensor-product transformation of the transformations $\lambda_1, \ldots, \lambda_r$. We have seen that a tensor T^* in V_r^0 is alternating if and only if its image $\tau(\mathrm{T}^*)$ is an alternating linear transformation. Consequently τ maps the subspace A_r^0 of V_r^0 consisting of the alternating tensors isomorphically onto $A(V_0^r; R)$, the subspace of $(V_0^r)^*$ consisting of the alternating linear functionals.

The alternation mapping A^r is a linear transformation of V_r^0 onto A_r^0 whose kernel is N_r [where $\wedge^r(V^*) = V_r^0/N_r$]. Hence $\mathrm{A}^r\colon V_r^0 \to A_r^0$ induces an isomorphism θ of $\wedge^r(V^*)$ onto A_r^0 defined by

$$\theta(\mathrm{T}^* + N_r) = \mathrm{A}^r(\mathrm{T}^*)$$

for all tensors T^* in V_r^0. In particular,

$$\theta(\lambda_1 \wedge \cdots \wedge \lambda_r) = \mathrm{A}^r(\lambda_1 \otimes \cdots \otimes \lambda_r)$$

Furthermore, if $\nu\colon V_0^r \to \wedge^r V$ is the natural homomorphism, the mapping $\mu \to \nu \circ \mu$ of $L(\wedge^r V; R) = (\wedge^r V)^*$ onto $A(V_0^r; R)$ is an isomorphism, according to Theorem 10-13. Let ψ be the inverse of this isomorphism. Then the composition of the isomorphisms

$$\wedge^r(V^*) \xrightarrow{\theta} A_r{}^0 \xrightarrow{\tau} A(V_0{}^r; R) \xrightarrow{\psi} (\wedge^r V)^*$$

is the desired isomorphism, as we shall demonstrate.

Let $\lambda_1, \ldots, \lambda_r$ be elements of V^* and $\alpha_1, \ldots, \alpha_r$ be elements of V. Then

$$\xi = \theta(\lambda_1 \wedge \cdots \wedge \lambda_r) = \sum_{S_r} (\text{sign } \pi)\ \lambda_{\pi(1)} \otimes \cdots \otimes \lambda_{\pi(r)}$$

and, as we have seen in Section 10-6,

$$\tau \circ \xi(\alpha_1 \otimes \cdots \otimes \alpha_r) = \sum_{S_r} (\text{sign } \pi) \lambda_{\pi(1)}(\alpha_1) \cdots \lambda_{\pi(r)}(\alpha_r)$$
$$= \det\ (\lambda_i(\alpha_j))$$

Consequently

$$\psi \circ \tau \circ \xi(\alpha_1 \wedge \cdots \wedge \alpha_r) = \tau \circ \xi(\alpha_1 \otimes \cdots \otimes \alpha_r) = \det\ (\lambda_i(\alpha_j))$$

or
$$[\sigma(\lambda_1 \wedge \cdots \wedge \lambda_r)](\alpha_1 \wedge \cdots \wedge \alpha_r) = \det\ (\lambda_i(\alpha_j))$$

as was to be proved.

In accordance with this result, we may identify $\sigma(\lambda_1 \wedge \cdots \wedge \lambda_r)$ with $\lambda_1 \wedge \cdots \wedge \lambda_r$. That is, we may regard $\lambda_1 \wedge \cdots \wedge \lambda_r$ either as an element of $\wedge^r(V^*)$ or as the linear functional on $\wedge^r V$ for which

$$(\lambda_1 \wedge \cdots \wedge \lambda_r)(\alpha_1 \wedge \cdots \wedge \alpha_r) = \det\ (\lambda_i(\alpha_j))$$

An element of $(\wedge^r V)^*$ [now identified with $\wedge^r(V^*)$] will be called an *exterior r-form* on V.

EXERCISE

Using the notation of the previous exercises, prove that, if V is a finite-dimensional vector space and V^* is its dual space. there is an isomorphism σ of $\dagger^r(V^*)$ onto $(\dagger^r V)^*$ such that

$$[\sigma(\lambda_1 \dagger \cdots \dagger \lambda_r)](\alpha_1 \dagger \cdots \dagger \alpha_r) = \sum_{S_r} \lambda_{\pi(1)}(\alpha_1) \cdots \lambda_{\pi(r)}(\alpha_r)$$

10-9 APPLICATIONS TO VECTOR SPACE BUNDLES

Let $\mathfrak{A}^m$ and $\mathfrak{B}^n$ be vector space bundles with $X(\mathfrak{A}^m) = X(\mathfrak{B}^n)$. Let $\{U_\alpha, g_{\alpha\beta}\}$ and $\{U_\alpha, h_{\alpha\beta}\}$ be systems of coordinate transformations for $\mathfrak{A}^m$ and $\mathfrak{B}^n$, respectively, with respect to a common covering of the base manifold. If $x \in U_\alpha \cap U_\beta$, then $g_{\alpha\beta}(x) \in L(V^m; V^m)$ and $h_{\alpha\beta}(x) \in L(V^n; V^n)$, and so we may define $q_{\alpha\beta}(x) = g_{\alpha\beta}(x) \otimes h_{\alpha\beta}(x)$.

$q_{\alpha\beta}(x)$ is a nonsingular linear transformation of $V^m \otimes V^n$. Then $\{U_\alpha, q_{\alpha\beta}\}$ is a system of coordinate transformations for a vector space bundle $\mathfrak{C}^{mn}$ which will be called the *tensor product* of $\mathfrak{A}^m$ and $\mathfrak{B}^n$ and will be denoted by $\mathfrak{A}^m \otimes \mathfrak{B}^n$. The tensor product of vector space bundles is a commutative and associative operation if we identify equivalent bundles.

Let $\mathfrak{T}^n$ and $\mathfrak{T}^{*n}$ be the tangent bundle and the cotangent bundle, respectively, of a differentiable manifold M. Let $\mathcal{V}_s{}^r$ be the set of all cross sections of the vector space bundle

$$\underbrace{\mathfrak{T}^n \otimes \cdots \otimes \mathfrak{T}^n}_{r \text{ factors}} \otimes \underbrace{\mathfrak{T}^{*n} \otimes \cdots \otimes \mathfrak{T}^{*n}}_{s \text{ factors}}$$

Then $\mathcal{V}_s{}^r$ is a module over the ring of differentiable functions on M. The elements of $\mathcal{V}_s{}^r$ are called *tensor fields* of contravariant order r and covariant order s. In particular, $\mathcal{V}_0{}^0$ consists of the differentiable functions on M, $\mathcal{V}_0{}^1$ consists of the differentiable vector fields on M, and $\mathcal{V}_1{}^0$ consists of the differentiable forms on M. If $\mathrm{T}_s{}^r \in \mathcal{V}_s{}^r$ and $\mathrm{T}'_{s'}{}^{r'} \in \mathcal{V}_{s'}{}^{r'}$, we may form the product $\mathrm{T}_s{}^r\mathrm{T}'_{s'}{}^{r'}$ which is an element of $\mathcal{V}_{s+s'}^{r+r'}$.

Let $\mathfrak{B}^n$ be a vector space bundle. Let $\{U_\alpha, g_{\alpha\beta}\}$ be a system of coordinate transformations for $\mathfrak{B}^n$. If $x \in U_\alpha \cap U_\beta$, we may form $h_{\alpha\beta}(x) = \wedge^r g_{\alpha\beta}(x)$. $h_{\alpha\beta}$ is a nonsingular linear transformation of $\wedge^r V^n$. Then $\{U_\alpha, h_{\alpha\beta}\}$ is a system of coordinate transformations for a vector space bundle $\mathfrak{C}^q$ with $q = \binom{n}{r}$ which will be called the *r-fold exterior power* of $\mathfrak{B}^n$ and will be denoted by $\wedge^r\mathfrak{B}^n$.

Let $\mathfrak{T}^{*n}$ be the cotangent bundle of a differentiable manifold M. Let $\mathcal{W}^r$ be the set of all cross sections of $\wedge^r\mathfrak{T}^{*n}$. Then $\mathcal{W}^r$ is a module over the ring of differentiable functions on M. The elements of $\mathcal{W}^r$ are called *exterior r-forms* on M. In particular, $\mathcal{W}^0$ consists of the differentiable functions on M, and $\mathcal{W}^1$ consists of the differentiable forms on M. If $\Omega^r \in \mathcal{W}^r$ and $\Omega'^{r'} \in \mathcal{W}^{r'}$, we may form $\Omega^r \wedge \Omega'^{r'}$, which will be an exterior $(r + r')$-form. If $r > n$, then $\mathcal{W}^r = \{0\}$. Let $\mathcal{W}$ denote the direct sum

$$\mathcal{W}^0 \otimes \mathcal{W}^1 \otimes \cdots \otimes \mathcal{W}^n$$

If multiplication is defined in $\mathcal{W}$ in accordance with the rules for the exterior algebra, $\mathcal{W}$ becomes an algebra called the *algebra of exterior forms*. If $r \leqq n$, then $\mathcal{W}^r$ may be regarded as a subspace of $\mathcal{W}$.

Theorem 10-21. *If $\mathcal{W}$ is the algebra of exterior forms on a differentiable manifold M, then there is a unique linear transformation $d: \mathcal{W} \to \mathcal{W}$ which satisfies the following conditions:*

(i) *If* $f \in \mathcal{W}^0$, *then* df *is the differential of* f.

(ii) *If* d *is restricted to* $\mathcal{W}^r$, *then* d *is a linear transformation of* $\mathcal{W}^r$ *into* $\mathcal{W}^{r+1}$.

(iii) *If* $\Omega_1{}^r \in \mathcal{W}^r$ *and* $\Omega_2{}^s \in \mathcal{W}^s$, *then*

$$d(\Omega_1{}^r \wedge \Omega_2{}^s) = (d\Omega_1{}^r) \wedge \Omega_2{}^s + (-1)^r \Omega_1{}^r \wedge d\Omega_2{}^s$$

(iv) $d \circ d = 0$.

Proof. Let U be a coordinate neighborhood of M with the coordinates $(x_1, \ldots, x_n)$. The properties i, iii, and iv imply $d(dx_i) = 0$, $d(dx_i \wedge dx_j) = 0$, $d(dx_i \wedge dx_j \wedge dx_k) = 0, \ldots$. It follows that for $f \in \mathcal{W}^0$

$$d(f\, dx_i \wedge dx_j \wedge \cdots \wedge dx_k) = df \wedge dx_i \wedge \cdots \wedge dx_k$$
$$= \sum_{l=1}^{n} \frac{\partial f}{\partial x_l}\, dx_l \wedge dx_i \wedge \cdots \wedge dx_k$$

This means that, if the mapping d exists, it is given locally by the formula

$$d(\Sigma a_{i_1,\ldots,i_r}\, dx_{i_1} \wedge \cdots \wedge dx_{i_r}) = \Sigma\, da_{i_1,\ldots,i_r} \wedge dx_{i_1} \wedge \cdots \wedge dx_{i_r}$$

on the elements of $\mathcal{W}^r$. Since d is a linear mapping, it must therefore be unique on $\mathcal{W}$.

The existence of d follows by showing that the local formula which we have derived provides a well-defined mapping d which actually has the properties i to iv.

The mapping d is called *exterior differentiation.*

EXERCISES

1. Fill in the missing details in this section.

2. Let M be a differentiable manifold. Prove that there is a diffeomorphism of the principal bundle of M into $\mathcal{V}_1{}^1$ which carries fibers into fibers (see Theorem 10-7).

3. Define the analogue for tensor fields of the mapping $\zeta_j{}^i$ of Exercise 1 in Section 10-4. This mapping is called *contraction.*

4. Let $\langle\alpha, \beta\rangle_m$ be a Riemann metric on the differentiable manifold M. Prove that $\langle\alpha, \beta\rangle_m$ determines a tensor field on M of contravariant order 0 and covariant order 2.

5. According to Theorem 10-20, if $\mathfrak{T}^n$ and $\mathfrak{T}^{*n}$ are the tangent bundle and cotangent bundle, respectively, of a differentiable manifold M, the fiber over m of $\wedge^2\mathfrak{T}^{*n}$ is the dual space of the fiber over m of $\wedge^2\mathfrak{T}^n$. If χ_1 and χ_2 are differentiable vector fields on M, then $\chi_1 \wedge \chi_2$ is a cross section of $\wedge^2\mathfrak{T}^n$. Consequently, if ω is a 1-form on M, we may compute the value of the 2-form $d\omega$ at $\chi_1 \wedge \chi_2$. Prove the identity $d\omega(\chi_1 \wedge \chi_2) = \langle\chi_1, d\langle\chi_2, \omega\rangle\rangle - \langle\chi_2, d\langle\chi_1\, \omega\rangle\rangle - \langle[\chi_1, \chi_2], \omega\rangle$.

References

LIE GROUPS

Chevalley, C.: "Theory of Lie Groups," I, Princeton University Press, Princeton, N.J., 1946.

Cohn, P. M.: "Lie Groups," Cambridge University Press, London, 1957.

RIEMANNIAN GEOMETRY

Cartan, E. J.: "La Géométrie des espaces de Riemann," Gauthier-Villars, Paris, 1925.

———: "La Théorie des groupes finis et continus et la géométrie differentielle, traitées par la méthode du repère mobile," Gauthier-Villars, Paris, 1937.

Chern, S. S.: Differentiable Manifolds, notes, Department of Mathematics, University of Chicago, 1959.

Goldberg, S. I.: "Curvature and Homology," Academic Press Inc., New York, 1962.

Helgason, Sigudur: "Differential Geometry and Symmetric Spaces," Academic Press Inc., New York, 1962.

Lichnerowicz, A.: "Théorie globale des connexions et des groupes d'holonomie," Edizioni Cremonese, Rome, 1955.

———: "Géométrie des groupes de transformations," Dunod, Paris, 1958.

Nomizu, K.: "Lie Groups and Differential Geometry," Mathematical Society of Japan, 1956.

FIBER BUNDLES

Milnor, J.: Differential Topology, notes, Department of Mathematics, Princeton University, 1958.

———: The Theory of Characteristic Classes, notes, Department of Mathematics, Princeton University, 1958.

Steenrod, N. E.: "The Topology of Fibre Bundles," Princeton University Press, Princeton, N.J., 1951.

Index